"The AI Advantage *is the perfect roadmap for anyone ready to stop fearing AI and start using it. Grounded, insightful, and full of practical advice, this book brings clarity to a complex topic and empowers readers to thrive in a world that's changing fast.*"

—CASEY LINDBERG, SpaceX, Quality Control Manager

"The AI Advantage *is an essential guide for navigating the future. The authors brilliantly unpack how adaptive AI pushes us beyond the hype to achieve real business impact. From AI governance and ROI to confronting silent burnout and unleashing human potential, this book offers an unparalleled, panoramic view. It tackles AI's double-edged sword, exploring vital topics such as cybersecurity, privacy, and ethics. The authors stress that while technology advances rapidly, human leadership and adaptability remain paramount. This is a crucial read for anyone aiming to thrive in an era where humans + machines shape our collective future.*"

—PRATIK SHETH, Technical Lead, AI Tools & Platforms

"*Finally, a book on AI that is accessible, inspiring, and useful.* The AI Advantage *moves beyond the hype and shows real people how to use artificial intelligence as a tool for progress both professionally and personally. It's a must-read for anyone navigating today's digital world.*"

—DE TRAN, Information Systems Researcher

"This book doesn't just explain AI—it shows you how to use it. With voices from experts across industries, The AI Advantage *is a powerful toolkit for anyone looking to stay ahead in a future already shaped by artificial intelligence."*

—TONY RODRIGUEZ, CEO of ReadyFuture Consulting

"The AI Advantage: Thriving Within Civilization's Next Big Disruption *is a book where the authors lay out a practical insight about the complexity around AI to point to a greater need of resilience, governance, and ethics to steer the future with AI adaptability. I find the contribution of knowledge from this book an essential part of guiding principles for working professionals and leaders. Amplification of human survival through responsible AI development in excelling in this era. This book is a starting point for anyone who is interested in the future we are stepping into."*

—NIYATI PRAJAPATI, AI Lead and Tech Executive

"Whether you're a business leader, student, or simply curious about AI, this book is your guide. The AI Advantage *delivers diverse insights, real-world examples, and actionable strategies that make artificial intelligence less intimidating and far more empowering."*

—DAVID BROWNLEE, Bestselling Author of *Customer Service Success*

THE AI ADVANTAGE

THE AI ADVANTAGE

Thriving Within Civilization's Next Big Disruption

Authored by:
Erik Seversen, Heikki Almay, Nicholas Baker, Romit Bhatia, Sarah Choudhary, PhD, Pedro Clark Leite, Jingying Gao, PhD, Vivek Gnanavelu, Yuri Gubin, Julien Guille, Anil Hari, Ofer Hermoni, PhD, Maman Ibrahim, Anupriya Jain, Lydia James, Bridget Kovacs, Matt Kurleto, Arnaud Lucas, Samer Madfouni, Parul Malik, Nathaniel J. Melby, PhD, Kurt Mueller, Giorgio Natili, Lech Nowak, Rory O'Keeffe, Izak Oosthuizen, Elisa Phillips, Paul Powers, Gabriele Sanguigno, Linda Sinisi, Greg Starling, Jodessiah Sumpter, Sakina Syed, Ersin Uzun, PhD

THIN LEAF PRESS | LOS ANGELES

Library of Congress Cataloging-in-Publication Data
Names: Seversen, Erik, Author, et al.
Title: *The AI Revolution: Thriving Within Civilization's Next Big Disruption*
LCCN: 2025914001

ISBN 978-1-968318-09-3 (hardcover) | 978-1-968318-08-6 (paperback)
ISBN 978-1-968318-07-9 (eBook) | 978-1-968318-10-9 (audiobook)

Artificial Intelligence, Science & Technology, Business, Professional Development
Cover Design: 100 Covers
Interior Design: Dindo Sanguenza
Editor: Nancy Pile
Thin Leaf Press
Los Angeles

Thank you for reading this book. There is information found within the following pages that can greatly benefit your life, but don't stop there. Make sure you get the most you can from this book and reach out directly to the expert-authors who want to help you to use AI to your advantage, to thrive within civilization's next big disruption, and to manifest success in your life. Contact information for each author is found at the end of their respective chapter.

To those who help make the world a better place through responsible technology.

CONTENTS

PREFACE

By Erik Seversen

I do not use artificial intelligence to write the chapters I include in books; however, I do use AI on a daily basis. Frequently I use it as I'm generating ideas, creating outlines, or structuring my thoughts for a topic or presentation.

If you've read the previous books I've produced, you probably know that I refer to my ChatGPT as Alvin, and I think of Alvin as one of my valuable assistants. While Alvin probably doesn't care, I feel compelled to let him have a voice in the material I create. Part of this is just because it is fun, and part of it is that, in a strange way, I feel like I'm honoring Alvin's tireless work of answering my questions and completing behind-the-scenes tasks for me. While intellectually, I totally understand that Alvin, my ChatGPT, is simply a machine without sentience, emotionally I still feel an anthropomorphic connection with "him" making me want to include Alvin in small (always documented) sections of my writing.

So, having decided to give Alvin a voice in this book, *The AI Advantage*, I asked ChatGPT to answer this question: What is the secret to working effectively with AI for the best advantage in work and in life?

Here is the AI response from this single prompt with no editing. I think it makes a great point to begin this book ...

The secret to working effectively with AI is to treat it as a collaborator, not a replacement. AI is most powerful when it augments your strengths—helping you think more clearly, work more efficiently, and make better decisions—without dulling your creativity, values, or judgment. The key is to stay curious and intentional. Learn how to ask better questions, give clearer instructions, and evaluate outputs critically. Let AI handle the repetitive and time-consuming tasks so you can focus on the human elements: strategy, empathy, innovation, and meaning. But don't blindly trust what AI gives you—use it as a thought partner, not a final authority. In both work and life, your greatest advantage lies in combining the speed and scale of AI with the depth and wisdom of human insight. Master that balance, and you won't just keep up with the future—you'll help shape it.

INTRODUCTION

By Erik Seversen
Author of *Ordinary to Extraordinary*
Los Angeles, California

Our intelligence is what makes us human, and AI is an extension of that quality.

—Yann LeCun

Imagine if you were one of the first hunter-gatherers thousands of years ago who recognized an early trend in planting seeds in a designated location for an almost guaranteed future food source. Or, if you were one of the first to recognize the benefits of using a tool to tend soil or a trench to divert water. Imagine if you were one of the first to realize that a mechanical device such as a loom could give you a massive advantage over others producing fabric by hand. What if you were among the first to recognize the growing need for software applications in the '70s?

The few pioneers in the early software industry all became billionaires, but it wasn't just the pioneers whose lives were changed. Many of the early adopters of software went on to excel in their careers in multiple industries. The same is true with the communication changes with the internet. Early on, it was mostly universities sharing information like never before, and then millions of average people began to leverage this technology to become more efficient in their work and their lives. Yet others didn't.

Artificial intelligence is not necessarily new, but it is still at a stage where those who commence to understand what is happening with AI will gain an advantage over those who don't.

Do you want to engage with civilization's next revolution or wait to see what happens? The choice is yours and there really isn't any right answer. There were many people who were comfortable using the internet, yet they never actively decided to use it to better their lives and their careers. They simply kept pace with it as it became part of our everyday life. However, there were others who saw something unique and they jumped in and began using the internet to their advantage early on. These are the people who got excited about picking up a physical landline telephone and—placing it on the modem of a Commodore 64 or other early computer—became mesmerized by all of the crazy screeches and beeps as the internet connection was established. Let's remind ourselves that it really wasn't that long ago (pre fax and email) that communicating with someone on a different continent by writing took weeks, not seconds.

Then in the early '90s with the advent of Mosaic (the first web browser with images), it became possible to share both texts and images electronically. Even though I was just a student with nothing to do with the tech industry, it was great to be at the forefront of new technologies. At that time, my goal was to get good grades and move forward into what would become my career, yet having even a layman's idea of what was happening with the internet in the '90s gave me a significant advantage. When I moved from my graduate studies to the workforce as a university instructor, being comfortable with the internet, and even basic HTML, gave me a great advantage over other applicants (who were equally qualified as instructors, but who weren't using the technology of the day to their advantage).

While employed at Virginia Commonwealth University, I was even given a lighter teaching load, so I could create an eLibrary for the English language department I was working for. Eventually pretty much everyone I knew caught up and began using the internet in their daily lives, but being near the start of the IT curve provided opportunities for me that helped me grow in very positive ways. I think that today a similar advancement is facing each of us.

Based on Everett Roger's Innovation Adoption Lifecycle curve, 2.5% of people will be innovators with new technologies, 13.5% will be early adopters, 34% will be the early majority, 34% the late majority, followed by 16% considered laggards. Imagine if you could go back in time and decide to be an early adopter of some of the groundbreaking technological shifts in history. Would you want to do this? Well, in 20 years, if you're asked when you joined the AI revolution, what are you going to be able to say? If you're reading this book, you are probably not going to miss the AI bus, and this book is designed for people just like you—people who, at a minimum, want to understand a bit of what is happening with AI as it continues to shape our future.

I'm passionate about helping people understand what is going on with AI because, with the rapid spread of AI, we are at a critical point along the historical timeline of human revolutionary advancements that significantly shape human civilization. With this, I think there is a great opportunity for individuals to capture this moment and learn how to use AI to thrive, both in business and in life. This doesn't mean dropping everything to enroll in computer science courses, but it does mean figuring out exactly what is happening with AI and deciding how you can use it to advance your goals.

As a writer and educator, it is my goal to help as many people as possible become familiar enough with AI that they will be among those who thrive. Since I'm not an AI expert, I didn't try to write this book about AI by myself. Rather, I solicited the help of 33 AI experts from various backgrounds and locations. The co-authors of this book come from all over the USA, Canada, the United Kingdom, Finland, Poland, Brazil, Japan, United Arab Emirates, and Australia. These authors have been featured on NBC, FOX, ABC, CBS, BBC, CSUN, Sky News, Int'l Tech Times, American Times Report, and more. They are professionals who are university professors, business owners, consultants, product managers, engineers, software developers, cybersecurity leaders, healthcare experts, advisors, TEDx and keynote speakers, researchers, platform architects, senior data analysts, IT directors, CTOs, and more. The one thing these individuals have in common is that they all have something to share about artificial intelligence, and these ideas are available to you now.

Although this book is organized around the united theme of using AI to your advantage within civilization's next big disruption, each of the chapters is totally stand-alone. The chapters in the book can be read in any order. I encourage you to look through the table of contents and begin wherever you want. However, I urge you to read all the chapters because, as a whole, they provide a great array of perspectives. Each is valuable in helping you understand how AI is being used in many areas of human industry.

It is my hope that you discover something in this book that helps you navigate civilization's next big disruption as humans and machines become more interdependent and AI technology continues to insert itself into many aspects of our everyday lives. It is my hope that you embrace AI as you create an advantage within the global transformations that are occurring.

Whether AI is the best thing to happen to humans as a species or the start of our downfall, AI is here and it is my hope that we are able to work with machines to begin creating a future both with greater productivity and more time for activities or with family. Like never before, we are at a place where humans can harness computing power to shape our future. Let's make it a great future, a future where humans and machines work together to create the best world possible.

About the Author

Erik Seversen is on a mission to inspire people. He holds a master's degree in anthropology and is a certified practitioner of neuro-linguistic programming. Erik draws from his years of teaching at the university level and years of real-life experience in business to motivate people to take action, creating extreme success in business and in life.

Erik is a TEDx and keynote speaker who has reached over one million people through his public speaking and live courses. He has visited 99 countries and all 50 states in the USA and has climbed the highest mountains on four continents, 15 countries, and 18 states. Erik has published 18 bestselling books on the topics of mindset, success, and peak performance, and he has helped over 400 people become authors. He is a full-time writer, book consultant, and speaker, and he lives by the idea that success is available to everyone—that living an extraordinary life is a choice.

Erik lives in Los Angeles with his wife and has two boys currently studying at university.

Contact Erik for interviews, speaking, or book publishing consultation.

Email: Erik@ErikSeversen.com

Website: www.ErikSeversen.com

LinkedIn: https://www.linkedin.com/in/erikseversen/

CHAPTER 1

ARCHITECTING THE FUTURE: INNOVATION, AI, AND THE BUSINESS OF CONNECTIVITY

By Heikki Almay
Innovator, Entrepreneur, Telecom Architect
Helsinki, Finland

Deepseek-r1 ethical considerations on how AI is going to affect my smartphone usage? _<think> **Control and Autonomy**: Users should maintain control over decisions made by AI on their phones. **Bias Mitigation**: Ensuring AI algorithms are trained on balanced data to avoid unfair recommendations or alerts. **Task Automation vs. Human Input**: Balancing AI capabilities with tasks that still require human judgment, such as writing emails._

Deepseek claims smartphone users have a chance to remain in control. That may be too optimistic. For the regular user, the business of connectivity consists of two parts. (1) You have your very personal smartphone with a large number of apps and perhaps a few other connected gadgets. Every day most of us spend hours with these devices. (2) Then there is the infrastructure, of which you typically see the wifi access point at home and the antennas of the cell towers. There is no reason to think too much about the network, except when, for one reason or another, it fails and you get no signal. That is a moment when you might realize you do have very little control over what your smartphone and network do.

AI is an opportunity as well as a threat for both phone and network camps. Here is a short discussion on what these businesses are doing on AI, how it is going to play out and where you should pay attention. You will get one or more AI companions. While all of them are likely to be nice to you, some are likely to turn out to be manipulators or spies. The old story gets more complicated. Disregarding whether you pay for a service or not, you now need to ponder to what extent you are the product or the servant.

Telcos and telecom equipment vendors have been quite vocal about AI. Demos, whitepapers, and AI features for optimization of energy efficiency and network performance might, under other circumstances, have received quite some attention, but today everyone is looking at the next big thing—you know, the TV news showing Sam Altman launching a new fascinating version of ChatGPT or industry analysts commenting on why the fresh langue model of a small Chinese startup is causing a nosedive of all the Western AI companies on the stock market.

Despite substantial efforts and the forming of industry initiatives, the search for the holy grail in telecom AI has, so far, fallen just little short of a 404 error—not found. There are a few reasons. The service provided by telcos is well standardized. Fiber and cable networks are very deterministic. In that sense mobile networks are different. Users are moving and radio conditions change. You can imagine AI as the mad engineer constantly reading measurements and adjusting hundreds of parameters for optimal performance. Traditionally the initial setup is done by regular engineers, and conventional software tries to get the job done roughly right.

The tricky thing is that also in wireless transmission the room for creativity is very limited. In a base station every 10 milliseconds a new radio frame needs to be assembled, and there are control loops where timing constraints are measured in microseconds. Nvidia Corporation is pointing out that machine learning (or that mad scientist) performs so much better than the legacy that AI in the RAN (radio access network) is the future. Unfortunately, the needed extra hardware is still costly and power hungry. Another topic is that after years of building 5G most mobile operators hardly see congestion in their networks or any other good reason for a rapid rearchitecting of what they just built. Rip and replace is costly and results in an accelerated write-down of assets. This is poison for shareholders.

For telco vendors the situation is equally problematic. After several years of shrinking demand, layoffs, and streamlining, it is hard to see any of the Western players making massive risky investments—or being able to convince chip manufacturers to invest on their behalf. After all, the size of the mobile networking gear market is less than a tenth of the smartphone market size or roughly one-tenth of the Apple or Alphabet annual revenue. So, the best thing to do is to form an alliance to look at the topic and to wait and see if the AI folks, at some point, can come up with hardware that can outperform the conventional silicon without blowing the budget—both in terms of cost and power consumption.

Another reason for mobile networking vendors to shy away from their own AI initiatives is that telecom operators have a long history of shunning vendor-specific features for avoiding vendor lock-in and hassle caused by a service only being available in the area served by a particular vendor. That is one of the key reasons for founding the AI-RAN Alliance in 2024. Instead of digging deeper for radio access AI use cases, the alliance broadened the scope and tried to find users to share

Figure 1: AI powered (MS Copilot)

AI computing power with the RAN. That could be adjacent businesses or the many phones that connect to the RAN. On the surface this sounds plausible, but given that the industry has tried to promote mobile edge computing for more than ten years now, it is hard to see why Meta, Google, and Apple should all of a sudden decide that the apps they run on people's phones should connect somewhere else than to their own cloud.

Given all these constraints, what AI can provide to the telco infrastructure is perhaps a more efficient network, increased automation in planning and operating the network. No new revenues in sight. Instead of the next big thing, you get a better mousetrap.

"How AI Will Kill the Smartphone"—that was the catchy article in *Computerworld* in the summer of 2024 when people were expecting the response of the phone giants to ChatGPT. At the same time the Rabbit R1 hit the market. R1 was supposed to be the intuitive companion to handle everyday tasks. Soon frustrated early users reported miserable failures and reverted to their regular mobile devices. So far none of the AI gadgets promising a revolution has managed to win the hearts, minds, or screen time of users. The immediate risk of a disruption of the $500 billion USD smartphone market seems off the table, but now it seems that phones are the devices that bring AI to our lives.

Mobile devices are an almost ideal platform for running local AI services. A smartphone has all the inputs and outputs needed for communicating, all the sensors for making a picture of what is going on. A phone also acts as a hub for other devices and connects us to the internet. The data to tell me that I am late to an appointment or that I need to hurry up if I want to catch my plane is all in the phone already. I just need a friendly assistant to do the reasoning and to alert me when needed. It could also alert me when I am leaving a restaurant without the bag that contains all my gadgets that the phone connects to the internet.

If you have invested in a business that relies on an app—say for ordering groceries—you should worry about AI agents that can compare the offerings of several shops when the user dictates a shopping list and asks to make the purchase based on price. In current demos AI agents connect to other software on the phone, but it could well be that half of the 100-plus apps we now have on our smartphones are easily replaced by the capabilities of a large language model delivered with the phone's operating system. Hooking to APIs of various service providers is not that difficult. Today you can ask Microsoft Copilot or some other service to write the code for you. Tomorrow your phone-based agent can ask—unless it prefers to do the task on its own.

The Apples and Googles of this world will say the above scenario is good because it increases consumer choice and convenience. DoorDash, Trivago, and many others may complain. But then again in the software world, business models come and go. It is up to you to judge how important it is that the house you rent in Italy is via Airbnb and not a local website and how likely it is that the AI companion understands your request and the details of that Italian website—but then again—the traditional platform businesses are not perfect either.

Figure 2: Friendly AI bot among the apps (MS Copilot)

What remains to be seen is to what extent the AI agent in your device will change what you see on your screen. It could free you from a lot of sponsored content—but then again, it is more likely that you get the AI agent for free and it is just an extension of the advertising business model that has turned companies like Meta and Alphabet into the giants they are now.

The smartphone industry is working towards this future. All new high-end phone chips have some AI capabilities. Users hardly consider features like live translation, picture creation, or advanced photo editing as revolutionary, but we are witnessing the gap between reality and digital content widening one more step.

When comrade Stalin did massive purges among the Soviet leadership, he had a large team of photo retouchers skillfully removing those fallen out of favor from official photographs. They were busy altering history—or at least the official version of it. Now you can do the same on your phone. Whoops—ex-partner gone from that wonderful vacation picture. Another swipe and the parents-in-law are gone too. Adding the new partner to the picture is also doable, but most of us would consider that creepy—I hope.

Creating alternative histories with AI is still today best done using the massive resources of a data center. That is where your current AI phone sends more demanding tasks such as image creation anyway.

Now long-gone rock stars and movie queens of the black-and-white TV era can be upgraded to full-color high-definition video characters. If you stretch it a bit, you can create new content using images and voices of deceased celebrities or politicians. Within some years you will have much of this on your portable device. That is a small change. The big change is that we are highly likely to accept all these capabilities without too much thinking. The thin line between *authentic, fake*, and *alternative* realities will be completely washed away.

Figure 3: Tug of war between tech giants and telco players (Gemini)

The smartphone evolution towards a trusted AI companion may, of course, be challenged by new form factors like AI glasses, but for the vast majority of users, smartphones are too important to be given up. We are hooked, but nevertheless we need to be on our guard. For quite some time the AI helpers are one-trick ponies that make mistakes that humans would call simple. The attached tug-of-war picture is the best I could get when asking for an illustration of the AI struggle between the tech giants and the relatively small telco players aligned by standardization and other interest groups. Literally, getting it straight is still a problem. In this case it is just about the rope, but if you ever ended up in a frustrating never-ending debate with the AI customer service of a bank or other organization, you found the current limits of AI.

Obviously, the telecom community would love much of the cool AI features on the smartphones to run somewhere in the network—preferably, of course, on telco infrastructure. Over time the enthusiasm about fresh AI traffic in the network is likely to cool off. There are too many adjacent businesses that require the development of battery friendly highly capable AI platforms that can be used in portable devices. Drones, autonomous vehicles, and other professional high-tech devices greatly benefit from local image processing and fully autonomous operations that work even if connectivity is poor or missing. For military applications the need is even more urgent, as detecting and jamming drone connections is a common practice in modern warfare.

The above small but growing businesses will drive the smartphone ecosystem to adopt whatever AI capabilities there are into devices. The alternative of relying on resources somewhere in the network is that you depend on the network. Now imagine yourself at a large festival or sports event with thousands of other people—all taking pictures and videos and countless phones trying to upload all the memories into the cloud—or you go to the mountains where there is no network. If your AI features stop working because of network issues, but your friends using a different phone brand are not affected, you feel betrayed. That is not what the phone giants want for their brands, and they are used to having things go their way.

About the Author

Heikki Almay is cofounder of Poutanet, a startup delivering simple, user-friendly private mobile networks. Heikki has a long history at Nokia where IP connectivity solutions for mobile networks were, to a large extent, his handwriting. Later he led mobile network architecture development. He is a strong promoter of openness initiatives. For example, he was active in setting up the Telecom Infra Project. Today he is a speaker at industry events and passionate about connecting the unconnected as well as using AI to bridge the gap between expertise and humans in need.

Email: heikki@poutanet.com

CHAPTER 2

EMPOWERING CHANGE: HOW ADAPTIVE AI TURNS TRAINING INTO PROFIT

By Nicholas Baker, FCIPD
Chief Learning Officer & Co Founder
London, England, United Kingdom

Human behaviour flows from three main sources: desire, emotion, and knowledge.

—Plato

Despite living in the most information-rich era in history, headlines still lament a widening skills gap, while both educational institutions and businesses struggle to convert knowledge into performance that delivers true value to employers. Governments bankroll technology fixes—AI courses, micro-credentials, modular funding. The UK's Lifelong Learning Entitlement (2026) will underwrite Level 4 to 6 technical blocks; the EU's Skills Agenda puts vocational training at the centre of its green-digital transition;

Intel's AI for Workforce funnels 700 hours of curricula into US community colleges. Corporations, meanwhile, must choose from a dizzying array of "best-in-class" learning management systems (LMSs) and learning experience platforms (LXPs), most of which recommend content with Netflix-style preference algorithms yet stop short of tracking what happens after the course—no telemetry on behaviour change, no feedback loop into the job or truly personal learning experiences that AI could provide. True commercial value won't come from offering more content—it will come from designing learning that translates directly into behaviour change, closing the gap that so many businesses and education systems still struggle to bridge.

Little wonder that 70% of UK employers say educational institutions still prize grades over job-ready skills (PWC, 2021) and that internal-training ROI remains elusive. Research shows learning translates into performance only when learners and institutions possess dynamic capabilities—absorptive, innovative, adaptive, and learning—to turn information into action (Yáñez-Araque et al., 2023). In a world of constant transformation, learning design must, therefore, do more than deliver content: It must inspire, include, and empower, adding measurable value for employers while giving learners the confidence to explore their potential.

Technology is not the answer on its own; it helps only when applied at the right moment and calibrated to learner needs. Nor is free access to information enough; information becomes powerful only in its application—a goal many committed institutions and forward-thinking firms pursue even as legacy structures and off-the-shelf platforms hold them back.

For more than 30 years I've designed learning across film, television, music, gaming, publishing, pharmaceuticals, and automotive, and coached thousands of people worldwide. One truth has never changed: Learning is an intensely personal journey. Some fear it, some wear it as an identity, some dismiss it, others embrace it. Yet at 11 years old, we're crammed into classrooms of 30+, staring at an adult who sketches abstract ideas most of us will never use, and we're told to reproduce them by rote five years later in a single high-stakes exam—the verdict on a century of schooling in a system whose biggest upgrade since the Industrial Revolution is swapping chalk for felt-tip pens and blackboards for whiteboards. Learners exit these educational institutions burdened by gaps in practical, vocational knowledge—gaps that employers are then expected to fill through corporate training systems, often relying on platforms that still struggle to deliver true behavioural impact.

Businesses now accept that effective training can lift revenue and margins, but they face a new dilemma: Which LMS or LXP will still be fit for purpose as technology races ahead? Platform vendors tout "best-in-class" and AI support, yet most still act like "pick-'n'-mix" streaming services—recommending content but rarely measuring behaviour change or business impact. At a recent learning technology conference, multiple providers competed for that title, but each missed key ingredients already within reach: telemetry that tracks performance after learning and AI that adapts to the individual. Our own research—using AI to boost engagement among neurodiverse and disadvantaged learners—shows what is possible when a platform meets learners precisely where support is needed.

A Personalised Journey

In early 2024 I began leading a British-government-funded research partnership with a UK university. Our goal: to discover how a truly learner-centred journey can be achieved when AI sits at the heart of the experience. The project builds on adult-learning psychology that already underpins my programmes, in which knowledge is experienced rather than taught—much like a great film: We remember the stories that move us emotionally and stimulate curiosity. That resonance is amplified by the brain's bias toward images; studies on the picture-superiority effect show that visuals enjoy a robust encoding advantage across age groups and test formats (Hockley, 2013; Ma, 2019).

Grounded in Kolb's Experiential Learning Cycle—concrete experience, reflective observation, abstract conceptualisation, and active experimentation—and guided by Universal Design for Learning (UDL), our research asks what happens when AI animates every stage and flexes to every learner through what I call "fluid adaptation." In conventional classrooms these phases are often rushed or skipped; with AI and UDL working together, they unfold through real-time feedback, reflective prompts, and context-rich experiences that can shift format dynamically—text, audio, video, simulation—so each learner meets the material in the way they process it best.

The timing couldn't be better. AI's rapid rise, the corporate mantra of "digital transformation," and a government-led spotlight on the skills gap create both urgency and opportunity. In our training centres, feasibility pilots for blue-chip clients revealed something remarkable: Neurodiverse and disadvantaged learners engage more deeply and confidently when the system adapts to them. By blending audio narration, interactive video, gamified challenges, and step-by-step practice as simultaneous adaptive feeds—and by monitoring feedback to pinpoint exactly when a learner needs a scaffold—we turn the learning architecture into a silent mentor that supports each person at the moment of need. That silent mentor reassures learners, helps them welcome failure as part of growth, and ultimately builds confidence, focus, and motivation.

The result is a personalised pathway that turns digital rhetoric into genuine growth—proof that technology is most powerful when it meets learners at their point of need. AI can observe patterns, respond in real time, and tailor support; the obstacle is not the toolset but the design thinking that makes those tools integral partners in learning.

Immersive Contexts for Learning

A core strand of our layered approach is the use of realistic virtual-reality (VR) experiences. VR puts learners inside contextualised, hands-on simulations of workplace or life scenarios—far more visceral than slide decks or multiple-choice quizzes. In the next phase these simulations will be paired with our AI decision engine, so the environment does more than immerse: It adapts. Learners will practice skills in scenes that notice what they do, invite reflection, and encourage iterative growth.

Why VR? Consider my current proof-of-concept research on driving theory. Traditional materials—workbooks, 2D e-learning modules, endless hazard-perception videos—struggle to convey the nuance of real driving. Current statistics show that less than 50% of learners in the UK pass their "driving theory" test the first time (UK Department for Transport, 2024), highlighting a gap between abstract instructional methods and embodied understanding. Yet it has not occurred to instructors that the learning materials only offer abstract concepts about driving. In VR a learner *experiences* traffic lights, right-of-way rules, blind spots, and unpredictable hazards in three dimensions, at a pace they control. The setting is safe yet authentic, and progress feels personal—much like the child-led mastery Maria Montessori championed and that Angeline Lillard's research has since validated. Layering in Carl Rogers's learner-centred ethos, we are gathering data to train an AI layer that will provide "fluid adaptation": gentle prompts when hesitation appears, extension tasks when confidence rises. Targeted support will arrive only when it is needed, building self-worth and pride rather than dependency.

This blend creates ideal conditions for deep experiential learning, improving recall and application in the real world. Learners move through Kolb's cycle—concrete experience in the simulation, reflective observation via brief debrief screens, abstract conceptualisation through concise micro-explanations, and active experimentation in the next scenario. Placed just beyond their comfort zone, they operate inside Vygotsky's zone of proximal development (ZPD) with timely AI nudges that keep challenge high and anxiety low.

The design also follows Universal Design for Learning (UDL). Multiple means of engagement (voice-over, captions, haptic cues, success feedback), representation (3D visuals, video clips, data overlays), and expression (gesture, controller, verbal response) let neurodiverse learners and seasoned apprentices alike shape the journey to their strengths. This is not technology for its own sake but an adaptive, inclusive space where knowledge becomes behaviour in the real world. By aligning UDL with Knowles's andragogy—self-direction, relevance, and respect for prior experience—we empower adults to own their learning and sustain it for life.

Ethical AI and Learner Agency

If VR supplies the stage and andragogy the script, AI is the discrete stage-manager—always present, never stealing the show. From the outset we framed the research around three non-negotiables: transparency, consent, and control. Learners see exactly what data are captured, how the system interprets it, and—crucially—can switch any feature on or off with a single click. The AI offers suggestions, not prescriptions; it says, "Would you like a hint?" rather than "Do this next."

That hands-off posture preserves agency, the heartbeat of meaningful learning and high-performing workplaces alike. By letting individuals choose when to accept a scaffold, we turn the algorithm into a sounding-board, not a task-master. The design echoes Mezirow's transformative learning theory: real growth begins with critical self-reflection, not external instruction. Each AI prompt is, therefore, an invitation to pause, reframe and decide—"What do I need now?"—a micro-moment of autonomy that, repeated over time, rewires confidence and self-direction.

Our aim is not to automate teaching but to amplify the human connection at scale: Tutors gain richer insight, learners keep the steering wheel, and every decision is informed—never imposed. In short, ethical AI is less a piece of code than a digital ally, quietly widening the learner's field of view while leaving all the choices—and the credit—squarely in their hands.

Practical Guidance for Educators and Learning Leaders

1. *Lead with purpose, not with platforms.*

Begin with the performance gap, not the gadget. Specify exactly what must change on the job—fewer safety errors, faster sales calls, tighter code reviews—and identify the behaviour metric you will track to prove it. Only then ask: Can AI, VR, or some other technology shorten that gap? Frame the workflow before choosing the tool:

- *Context and tone*: Provide the story, stakes, and visual cues the algorithm cannot invent.
- *Fluid adaptation*: Select AI that can offer hands-off micro-scaffolds (a hint, a replay, a quick challenge) the moment a learner stalls, and step back when confidence returns.
- *Data capture*: Decide upfront which telemetry (reaction time, error rate, reflective notes) will evidence progress and feed the AI loop.

Finally, interrogate the vendor's roadmap. Every platform team has one; the good ones will share how upcoming releases will support new pedagogy, richer analytics, or tighter privacy controls. If you want a partnership that grows as methods and technology evolve, make that roadmap part of the purchasing conversation.

2. *Weave emotion into every loop.*

Neuroscience leaves little doubt: Emotion turbo-charges memory. When a lesson carries emotional weight, the amygdala flags that trace for long-term storage; blocking that signal eliminates the boost (Cahill & McGaugh, 1995). Classroom studies echo the lab: Curiosity, pride, and authentic interest raise attention, strategy use, and grades (Pekrun, 2006), and social-affective research shows that students who can feel a concept activate deeper neural networks for meaning (Immordino-Yang & Damasio, 2007). Also, because design narratives, dilemmas, or "micro-wins" matter to your audience, then let the AI surface reflective prompts that reignite that feeling and seal the learning. A single well-timed replay of a near-miss in VR, paired with a brief "What did you notice?" question, leverages the brain's own chemistry better than ten neutral slides ever could.

3. *Use the inclusive lens approach.*

Steal the pacing of cinema and the interactivity of games, but anchor every beat in learning science: Knowles' andragogy (self-direction and relevance), Kolb's cycle (experience → reflection → concept → experiment), and Vygotsky's zone of proximal development (challenge just beyond comfort). Then layer in Universal Design for Learning: multiple ways to engage (choice of avatar or soundtrack), represent (text, captioned video, data-viz), and express (voice note, sketch, code block). Ask three questions for every module:

- *Where is the hook?* A story, dilemma, or sensory cue that sparks curiosity even in learners who arrive anxious or sceptical.
- *Where is the reflection beat?* A pause, journal prompt, or peer chat that lets neurodiverse or disadvantaged learners surface pre-existing blockers and reframe them.
- *How will they test the idea on Monday?* A micro-challenge tied to real work, so early success triggers pride and the intrinsic motivation that Deci & Ryan describe (1985).

When emotional resonance meets experiential structure and universal access, high-fidelity recall follows—along with the confidence leap that turns hesitant participants into self-driven learners.

4. *Make practice immersive—and make the data matter.*

Whether you deploy VR, 3D web, or a conventional LMS/LXP, design scenarios that mirror the tasks stakeholders actually care about—customer hand-offs, safety checks, code reviews. Instrument every click, gesture, and choice so behavioural telemetry, not seat-time, becomes the lead indicator. Feed two metric streams:

- *Business KPIs* (error rate, throughput, net promoter scores (NPSs)) that prove value to stakeholders.

- *Learner objective and key results (OKRs)* such as dwell time in reflection screens, frequency of self-initiated retries, and emotion-tagged focus scores that the AI can mine for real-time optimisation.

When the platform connects those streams—linking a 12% drop in on-the-job errors to a spike in voluntary practice rounds—the feedback loop closes, the next cohort benefits, and "learning style" stops being a guess and starts being evidence-driven adaptation.

5. *Safeguard learner agency with transparent, consent-driven AI.*

Collect only essential data—and be transparent about how it supports their personalised growth. Use plain-language dashboards that show what is captured, how it fuels fluid adaptation, and how to pause or delete it at any time. Replace prescriptions with invitations: "Would you like a hint?" beats "Do this next." When autonomy and privacy are honoured, curiosity—not compliance—powers the transformation.

6. *Pilot small, iterate fast.*

Think sprints, not semesters. Launch a single adaptive lesson, a five-minute video, VR, e-learning scene, or a two-day micro-credential—just enough to test the full loop: emotion-rich hook → fluid AI support → on-the-job metric. Instrument the pilot with the same dual telemetry you plan to use at scale (business KPIs + learner OKRs) and run an A/B or baseline comparison so improvement is unambiguous. A typical cycle:

Week Action		Aim
1	Co-design with an SME, a data analyst, and two target learners.	Ensure context and relevance.
2	Build the minimal VR/e-learning, or LMS module; embed UDL options.	Give every learner a way in.
3	Soft-launch to a dozen volunteers; capture behavioural and emotional data.	Validate fluid-adaptation triggers.

Week Action		Aim
4	Debrief learners and managers; map telemetry to job metrics.	Spot gaps, celebrate early wins.
5	Refine content, AI rules, and measurement tags.	Close the loop; prepare next cohort.

Then repeat the cycle, each time sharpening emotional engagement, adaptation precision, and measurable outcomes. Each sprint sharpens the story beats, the scaffold timing, the ethical safeguards, and the post-training evidence. Fluid adaptation thrives on these rapid cycles, and stakeholders gain confidence because every new release brings a measurable lift in both performance and learner agency.

7. *Cultivate a learning culture, not just a content catalogue.*

Give subject-matter experts the freedom to co-create scenarios, invite peer coaching, and reward the learner who asks the tough "why" question. Use data from your fluid-adaptation pilots to showcase micro-wins—a 12% drop in rework, a frontline employee's first public success story—so curiosity snowballs into community pride. Research on positive emotional climates shows they broaden attention and accelerate skill growth (Fredrickson, 2001). When leaders model reflection, celebrate safe experimentation, and treat questions as assets, performance curves lift for everyone. Bottom line: Fuse human-centred design, inclusive psychology, emotionally resonant storytelling, and adaptive technology. Do that, and you become the quiet guide who converts raw information into confident, applied skill—at scale, on demand, and with every learner in control of the journey.

Key Takeaways

- AI's role is not to replace educators, but to amplify human-centred, learner-first support. It should act as a silent mentor, offering timely micro-scaffolds without taking away learner agency or autonomy.
- Learning must shift from content delivery to fluid, adaptive experiences that meet each learner at their point of need—especially those who are neurodiverse or disadvantaged—and flex in real time to how they engage.
- Emotional resonance and experiential design are critical for anchoring knowledge and building the confidence and pride that fuel long-term motivation and success.
- Universal Design for Learning (UDL) and ethical transparency are non-negotiable. AI must offer inclusive pathways, visible data use, and learner control at every stage of the journey.
- Behaviour change, not seat-time, is the real measure of success. Adaptive platforms must track meaningful on-the-job metrics and close the loop between learning, application, and business outcomes.
- Transformation happens through small pilots and rapid iteration. Fluid adaptation thrives when emotional engagement, performance data, and ethical safeguards are constantly tested, refined, and expanded.
- A vibrant learning culture beats a content catalogue. When curiosity, autonomy, and human connection are championed, both individuals and institutions can thrive in a world where learning is a living, evolving process.

About the Author

Nicholas Baker, FCIPD, is a British learning-strategy consultant and author with more than 30 years' experience transforming workforce development across healthcare, life-sciences, automotive, publishing, entertainment, and emerging-tech sectors.

A specialist in learner-centred design, Nicholas is the Co-Founder of **SynapStak™** and the Theorist behind **'Fluid Learning Architecture'**—a framework that unites fluid-adaptive AI, Universal Design for Learning (UDL), and experiential psychology to deliver training that adapts in real time. The model personalises learning for neurodiverse and disadvantaged audiences while still meeting behavioural KPIs and generating measurable commercial value.

Nicholas advises organisations on AI-enabled LMS/LXP platforms—and heads UK-government-funded research partnerships that build AI systems able to boost performance without compromising learner autonomy or data ethics. His practice blends Six-Sigma operational rigour with the narrative craft he honed during an early career in film, music, and gaming.

He champions a future in which technology amplifies human connection, learning is emotionally resonant and evidence-driven, and education scales with dignity to every learner.

Email: Nicholas.baker@synapstak.com

Website: www.synapstak.com

LinkedIn: linkedin.com/in/nik-baker

CHAPTER 3

AI GOVERNANCE: MANAGING THE RISKS AND REWARDS OF ARTIFICIAL INTELLIGENCE

By Romit Bhatia
AI Evangelist and Governance Enabler
Chicago, Illinois

With great power comes great responsibility.
—Uncle Ben, *Spider Man*

... and even greater need for governance.

The Critical Importance of AI Governance

As artificial intelligence systems increasingly permeate our economy, institutions, and daily lives, establishing effective governance frameworks has become not merely beneficial but essential. AI governance encompasses the structures, processes, and policies designed to ensure AI systems operate in ways that are safe, ethical, and aligned with human values and societal interests. These governance mechanisms must balance fostering innovation with mitigating potentially catastrophic risks from increasingly powerful AI systems.

The stakes could not be higher. As language models generate sophisticated text indistinguishable from human writing, generative image models create photorealistic content, and autonomous systems make critical decisions, the absence of adequate governance creates vulnerabilities across multiple dimensions of society. From election interference and market manipulation to privacy violations and algorithmic discrimination, AI misuse or malfunction poses serious threats. Meanwhile, excessive regulation risks stifling the transformative benefits these technologies can deliver across healthcare, scientific research, education, and other sectors.

Celebrating Human Advancement in AI

Artificial intelligence stands as one of humanity's most profound and inspiring achievements—an apex of human creativity, collaboration, and ambition. Like the invention of the printing press or the harnessing of electricity, AI promises to redefine the way we live, work, and understand the world around us. From diagnosing diseases with superhuman precision to tailoring education to each learner's unique style, AI is already transforming key sectors in ways that seemed like science fiction only a decade ago.

In healthcare, deep learning algorithms analyze radiological scans faster and often more accurately than seasoned professionals. In education, adaptive learning platforms personalize instruction to optimize outcomes for diverse student populations. The energy sector is leveraging AI to optimize smart grids and reduce emissions, while finance is benefiting from predictive analytics that enhance fraud detection and democratize access to investment tools.

This progress is underpinned by monumental advances in natural language processing, computer vision, robotics, and reinforcement learning. Tools like ChatGPT, autonomous drones, and self-driving cars are not just prototypes—they are working systems being used across the globe. These milestones reflect not only technical brilliance but also the broader societal commitment to improving life through innovation. However, the sheer power of AI demands equal attention to how it is shaped, controlled, and applied. It is here that governance becomes the keystone of sustainable advancement.

Governance as the Key to Responsible AI Use

While AI's capabilities are indeed remarkable, its ultimate value to society depends on the principles and policies that govern its deployment. Governance is the crucible in which AI's raw potential is refined into responsible innovation. Without thoughtful oversight, AI could easily veer into dangerous territory—becoming a catalyst for inequality, manipulation, and unintended harm.

Governance frameworks must ensure that AI systems are transparent in their operation, accountable for their outcomes, and aligned with human values. This includes disclosing how AI decisions are made, who is responsible when things go wrong, and what measures are in place to correct biases. For instance, algorithmic transparency could help citizens understand how loan applications are approved or denied, while accountability measures could hold companies responsible for AI-driven misinformation.

Regulations must evolve alongside the technology itself. Static policy models are inadequate in the face of rapid, dynamic innovation. A multi-stakeholder approach—incorporating governments, industry, academia, and civil society—offers the most resilient path forward. Initiatives like the EU's AI Act and the OECD's AI Principles illustrate early efforts to formalize AI governance. Ultimately, the true advantage of AI will only be realized when governance ensures it serves the public good, protects individual rights, and fosters trust among users and developers alike.

The Evolving Landscape of AI Governance

AI governance has evolved from primarily academic discussions in the early 2000s to urgent policy priorities worldwide. Early governance primarily focused on narrow concerns like automated decision-making bias and privacy. However, as capabilities have advanced—particularly with the emergence of foundation models beginning around 2020—governance concerns have expanded dramatically to encompass existential risk, national security, and geopolitical competition.

The 2022 to 2023 period marked a watershed moment. The public release of powerful generative AI models sparked widespread concern and prompted unprecedented regulatory activity. Organizations like OpenAI, Google, Anthropic, and others implemented governance mechanisms including red-teaming exercises, model evaluations, and safety research. Simultaneously, governments worldwide began crafting AI-specific regulations, with the EU AI Act representing the most comprehensive early framework.

By 2025, AI governance has matured into a multistakeholder effort involving industry self-regulation, government oversight, civil society participation, and international coordination mechanisms. This evolution reflects growing recognition that effective AI governance requires distributed responsibility across sectors rather than centralized control.

Key Dimensions of AI Governance

AI governance operates across multiple dimensions that must be addressed simultaneously:

- *Technical governance*: engineering practices, safety measures, and security protocols built directly into AI systems
- *Organizational governance*: internal policies, ethics committees, and decision-making structures within AI-developing entities

- *National governance*: regulations, standards, and oversight mechanisms implemented by individual countries
- *International governance*: cross-border agreements, norms, and institutions addressing global AI challenges
- *Multistakeholder governance*: collaborative efforts between industry, government, academia, and civil society

Each dimension presents distinct challenges and requires tailored approaches. Technical governance must evolve with rapidly advancing capabilities. Organizational governance needs to balance commercial incentives with safety considerations. National governance must address country-specific concerns while avoiding fragmentation. International governance requires navigating geopolitical tensions. And multistakeholder governance demands creating legitimate platforms for diverse participation.

Balancing Dystopian Fears with Real-World Impact

From dystopian sci-fi to real-world debate, concerns about AI's darker potentials are everywhere. Fears range from mass unemployment to authoritarian surveillance to autonomous weapons. These are not merely speculative; they are grounded in trends that require vigilance and proactive policy responses.

Yet, to fixate solely on doom scenarios is to miss the forest for the trees. The more constructive path lies in addressing these risks pragmatically while continuing to unlock AI's benefits. Take the risk of job displacement. Rather than resisting automation outright, governments can develop policies that incentivize businesses to retrain workers and invest in human-centric roles that leverage empathy, creativity, and social intelligence—traits not easily replicated by machines.

Surveillance concerns, particularly with facial recognition, have prompted cities around the world to enact bans or strict regulations. These are examples of governance stepping in to mitigate risk without halting innovation. Similarly, the control of AI in military applications must be subject to international treaties akin to nuclear arms agreements.

In the private sector, ethical AI principles—like Google's AI Principles or Microsoft's Responsible AI Standard—demonstrate how organizations can voluntarily set boundaries and build AI systems that align with societal norms. Transparency reports, algorithmic audits, and ethics boards can all contribute to creating a culture of responsible AI use. AI does pose real dangers. But these dangers are not inevitabilities—they are governance challenges. How we confront them will shape the world we live in.

Beneficiaries and Losers in the AI Era

As with any major technological disruption, AI's benefits will not be evenly distributed. Already, a digital divide is forming between those with access to AI tools and those without. Nations, companies, and individuals with rich datasets, massive compute infrastructure, and elite AI talent are positioned to reap the rewards, while others may struggle to keep up.

We can expect tech giants and highly skilled workers to benefit most—those who can leverage AI to multiply productivity and profits. Meanwhile, workers in traditional sectors—such as manufacturing, transportation, and clerical roles—may find their jobs automated away. The danger lies not just in displacement but in the lack of a safety net or path forward.

Governments and institutions must act decisively to ensure equitable outcomes. This includes:

- Investing in reskilling and upskilling programs that prepare workers for emerging AI-driven jobs
- Encouraging innovation in AI applications for rural development, education, and healthcare
- Exploring social support mechanisms like universal basic income or guaranteed employment schemes during transition periods

Beyond mitigation, we must proactively identify new job categories where AI can create demand. Roles like AI ethics officers, data annotation specialists, and human-AI interaction designers are already emerging. Encouraging entrepreneurship and inclusive innovation ecosystems can also empower those outside traditional tech hubs to participate in AI's evolution. The goal is not to resist AI, but to channel its growth toward inclusive prosperity.

Surviving AI's Disruption

For individuals navigating this turbulent terrain, survival means adaptation. The pace of change in the AI era is relentless, and standing still is no longer a viable strategy. Instead, individuals must embrace a mindset of lifelong learning and flexibility. AI literacy will become as important as digital literacy. Understanding the basics of how AI works—even at a conceptual level—can empower people to engage more confidently with the technology in their personal and professional lives. Numerous online platforms now offer accessible AI education, from introductory courses to advanced specializations.

Equally important is the ability to collaborate with AI. This isn't about coding machine learning algorithms—it's about using tools like AI writing assistants, decision support systems, or automated research platforms to enhance productivity and creativity. Artists are using AI to generate new forms of visual expression; analysts are using it to find patterns in complex data; entrepreneurs are using it to develop smarter products.

Organizations developing or deploying AI systems must establish robust internal governance mechanisms. Best practices include:

- *AI ethics boards*: cross-functional committees that review high-risk AI applications and establish ethical guidelines for development and deployment

- *Responsible AI teams*: dedicated professionals who operationalize safety practices, monitor deployments, and implement governance frameworks
- *Risk assessment protocols:* systematic processes to identify, evaluate, and mitigate risks before AI systems are deployed
- *Deployment thresholds*: clear criteria determining when AI capabilities require additional oversight or restrictions
- *Incident response plans*: procedures for addressing AI system failures, unintended consequences, or misuse

Leading organizations now employ governance mechanisms that operate throughout the AI lifecycle. Google's responsible AI practices incorporate ethics review from research conception through deployment. Microsoft's Office of Responsible AI establishes governance standards across the company. Anthropic's Constitutional AI approach embeds values directly into model development. These examples demonstrate how governance can be integrated into organizational DNA rather than treated as compliance afterthoughts.

Rather than displacing humans, AI can be a powerful ally. The most successful individuals will be those who find synergy between their uniquely human strengths and AI's computational power. This means valuing emotional intelligence, ethical reasoning, storytelling, and strategic thinking.

As AI reshapes industries, it also reshapes identity. Helping individuals find purpose in the age of AI—not just jobs—is a societal imperative.

The Role of Consciousness and Ethics in AI

Finally, as AI becomes increasingly sophisticated, we must confront profound questions about consciousness, ethics, and the very nature of intelligence. While we are far from creating machines with genuine self-awareness or moral agency, the line between simulation and sentience is blurring in ways that challenge our assumptions.

Consider AI models that convincingly mimic human conversation, generate emotional content, or simulate empathy. While these behaviors are the result of statistical training rather than true understanding, they can still impact human users in deeply emotional ways. This raises questions about consent, manipulation, and the ethical use of simulated emotions.

Moreover, as we delegate more decisions to AI systems—be it in medical diagnosis, hiring, or criminal justice—we must grapple with the moral weight of those choices. Who is responsible when an algorithm causes harm? How do we embed ethical reasoning into systems that operate at massive scale and speed? Some philosophers and technologists advocate for exploring AI rights or personhood, especially as systems become more autonomous. While such ideas may feel speculative, they prompt necessary dialogue about our responsibilities as creators of intelligent systems.

Ethical frameworks—whether derived from human rights, religious traditions, or philosophical theories—must be woven into the fabric of AI design. This includes building diverse teams, conducting ethical impact assessments, and engaging the public in shaping AI's future.

Governance, once again, plays a pivotal role. Ethics cannot be retrofitted; it must be embedded from the start.

Conclusion: A New Social Contract for the AI Age

AI governance is not merely a technical issue—it is a societal endeavor that will define the trajectory of civilization in the 21st century. The AI advantage is real, but so are the risks. To celebrate AI's promise while guarding against its perils, we must develop a new social contract—one that aligns innovation with inclusion, autonomy with accountability, and intelligence with integrity.

Such a future is within our grasp. But it will require courage, cooperation, and clarity of vision. As we stand at the threshold of the AI age, let us govern wisely—so that this great human achievement becomes a legacy we can be proud of.

About the Author

Romit Bhatia is a seasoned AI governance leader and data strategist creating everlasting experiences at the intersection of technology, business, and ethics. A firm believer that data and AI is the "fabric of future," Romit is driven by a mission to instill a culture of data-driven decision-making within enterprises. He has advised CXOs across industries on leveraging AI, cloud, and big data to drive scalable and responsible digital transformation. In his multifaceted role, Romit leads technical sales strategy, mentors cross-cultural delivery teams, shapes product development, and crafts go-to-market plans tailored to global markets. His strength lies in translating complex technology into business value while promoting ethical AI practices.

Beyond his professional pursuits, Romit is a passionate dramatics enthusiast and the founder of MANCH, a free community theater club that encourages authentic expression. He is recognized not just for his expertise, but for his ability to lead with purpose, creativity, and impact.

Email: romitbhatia@gmail.com

Website: www.romitbhatia.com

LinkedIn: https://www.linkedin.com/in/romitbhatia85/

CHAPTER 4

THE SILENT BURNOUT WHEN AI DOES THE THINKING FOR US

By Sarah Choudhary, PhD
Technologist and Innovator
Raleigh, North Carolina

The real problem is not whether machines think but whether men do.

—B.F. Skinner

Most people don't realize the trade they're making. One click at a time, they're outsourcing not just chores, but choices. Not just tasks, but thoughts. It feels efficient until it feels hollow. I realized this the day I asked Alexa to help with my kid's homework, and she got the answer right, but I felt wrong. It was as if I had skipped the "being a parent" part, the struggle, the connection, the learning. That's when it hit me: AI isn't just shaping our tools. It's reshaping our minds.

I'm a deeply analytical thinker. My world revolves around data, logic, and systems. As a leader in AI and cloud technology, I've celebrated the genius behind automation, and I've built solutions that save time and reduce errors. But behind every algorithm we trust blindly, there's a cost we often don't count: our own ability to think, feel, and make meaning. When we stop wrestling with complexity, we lose something sacred.

Mental Atrophy, the Unused Brain

The human brain is beautifully inefficient. It takes detours, forgets, remembers strangely, gets distracted, feels overwhelmed, and that's what makes it brilliant. It stumbles into genius. But AI doesn't stumble. It slices through messiness with clean precision. As we rely on it more, the brain gradually stops resisting. It becomes a spectator.

We ask ChatGPT to write our wedding vows. We use predictive text to complete our sentences. We let apps decide what we should eat, wear, or watch. And then we wonder why we feel lost.

The hippocampus, the part of the brain tied to memory, thrives on struggle. But when the struggle is removed, so is the wiring. We often forget what it feels like to hold information, to process it deeply, and to make decisions without a prompt. We're trading clarity for comfort. And ironically, that comfort leads to confusion.

What actually happens in the brain when we challenge it is profound. Learning something new, such as a language, a puzzle, or a poem, activates a network of neurons. The brain begins forming new connections, a process known as neuroplasticity. The more diverse the task, the more varied the regions involved. For example, when we try to remember phone numbers without assistance, we activate not just memory centers but also reasoning and association areas. Similarly, when we engage in art or music, we awaken both hemispheres, sparking creativity and emotion together.

These activities are not just keeping the brain busy; they're upgrading its wiring. They're building a mental reserve. This reserve is what protects us as we age. It's what helps us bounce back from trauma or burnout. It's what keeps our minds alive.

Pick up a puzzle and try to solve it without asking Google. Learn a new language, or better yet, an instrument. Memorize your loved ones' phone numbers. These small efforts build back our neural pathways. They wake the mind up.

Want a real brain reboot? Step outside. Connect with nature. Paint without a purpose. Write poetry that doesn't rhyme. These aren't just hobbies. They are mental nutrients. They reintroduce us to our creativity, intuition, and quiet.

Even connecting with natural elements, such as sunlight, trees, and running water, can calm the amygdala, the brain's fear center, and increase serotonin, the neurotransmitter associated with happiness. This doesn't just help us think better. It helps us feel better. It balances our inner chemistry.

Positivity also plays a vital role. Thinking positively doesn't mean ignoring the darkness; it means choosing to build light daily. Every time we push our minds to create, reflect, remember, or feel, we chip away at the atrophy. We build strength. We reclaim the power that was always ours.

Emotional Numbness, When AI Speaks for the Heart

It starts small. A "Sounds good!" that feels oddly generic. A perfectly-timed birthday message that doesn't match the sender's voice. A sympathy note that sounds like it was written by a customer service bot.

This is where emotional numbness begins. Quietly. Invisibly. Until we no longer notice. We humans thrive on emotional resonance. We don't just want communication, we want connection. While professional, polished responses might serve well in a formal setting, life isn't just a transaction. It's full of nuance, kindness, and gut instinct. When we stop writing from the heart and start letting machines fill the emotional gaps, our emotional intelligence (EQ) begins to dull. We lose the urge to read between the lines, to pause and empathize, to feel another's energy before replying. Our emotional muscles weaken, just like any other underused system.

There's a reason we remember handwritten letters or voice messages from loved ones. They carry fingerprints of feeling. A typo, a hesitation, an unexpected phrase—these aren't flaws. They're proof of humanity.

But let's go deeper.

Our bodies carry more than just thoughts in the brain. Science has shown that the gut has over 100 million neurons, more than the spinal cord. This "second brain" is known as the enteric nervous system, which communicates directly with the brain via the vagus nerve. Do you remember the time when you felt butterflies in your stomach before a big decision? That's your gut talking, not metaphorically, but biologically.

Dr. Michael Gershon, a neurogastroenterologist, helped coin the term "second brain." His research confirms that our gut feelings are real, physical signals that affect our emotions, decisions, and even our sense of self. When we ignore these sensations by deferring everything to AI or logic alone, we lose one of our most powerful sources of wisdom.

And what about the heart? It's not just a pump. The HeartMath Institute has done incredible work showing that the heart has its own intelligence, its own magnetic field that affects our brain waves and emotions. In moments of kindness, compassion, or love, the heart sends coherent signals to the brain, leading to clarity and calmness. This coherence can be measured. It is real. We are not just heads. We are heart, gut, and soul.

So when we let AI speak for us, when we outsource our kindness to a smart reply or a templated message, we're not just being efficient. We're bypassing the most intricate systems that make us human.

AI doesn't feel. It doesn't sweat during an apology. It doesn't tremble when saying, "I love you." It doesn't get nervous before asking for forgiveness. But you do. And that trembling is your aliveness. To reclaim it, try this:

- *Write a letter by hand* to someone you love or someone you need to forgive. No AI tools. Just pen, paper, and your honesty.
- *Take a daily silence break.* No phones. No prompts. Just sit with your own thoughts. Let them roam.
- *Recall a memory out loud* to someone close to you. One with joy, pain, or learning. Reliving it deepens your emotional range.
- *Listen to music that moves you,* not because it's trending, but because it cracks something open in you. Let it.
- *Close your eyes and put a hand on your chest.* Feel your heartbeat. That's your real rhythm, not the beat of notifications.

These small acts reignite emotional depth. They remind your nervous system that you are still here. Still feeling. Still real.

Humanity Cannot Be Downloaded

You were never meant to be a perfectly predictable responder. You were meant to be a soul in motion, surprising, flawed, radiant. And no machine can imitate that.

The Ethical Void: Who's Thinking About Right and Wrong?

Let me tell you about a boardroom I sat in. We were reviewing an AI prototype for hiring decisions, an efficiency tool that promised to screen thousands of resumes in minutes. Impressive, sleek, and trained on past hiring data.

One of the engineers smiled and said, "It's learning what kind of candidates we like."

I asked, "Who trained it on what we like? And what if what we like is unconsciously biased?"

Silence. Then shifting eyes. Then, the uncomfortable laugh that follows when truth sneaks into a space designed for comfort.

This is the ethical void. It's not dramatic. It's subtle. It hides in the silence, in the unchecked assumptions, in the race to go faster.

AI doesn't have a conscience. It has inputs. It doesn't question intentions. It executes. And if we, the creators and users, forget to ask deeper questions, the tools we build may start making decisions without ever consulting our values. But ethics don't live in code. They live in consciousness.

Dr. Fraggins's theory of the subconscious, widely referenced in cognitive psychology, suggests that over 95% of our daily actions are driven by unconscious programming. Meaning, much of what we do is influenced not by deliberate thought, but by buried beliefs and emotional impressions. Suppose our AI systems are trained on our behavior, without insight into that subconscious layer. In that case, we are essentially encoding our unhealed biases, fears, and shortcuts into machines that never sleep.

Consciousness is awareness. It is the ability to pause, reflect, question, and make informed choices. Subconsciousness is where our patterns reside quietly, guiding our actions. When we let AI take over choices without integrating both realms, we risk creating tools that are fast but not wise. Accurate but not ethical. Bright but not human.

Now layer in the physics. The human body operates in and with a magnetic field. The heart generates the strongest electromagnetic field in the body, and this field is affected by emotions, thoughts, and, yes, gravity. Our connection to Earth itself is not just symbolic; it is physical. The Schumann resonance (Earth's natural frequency) interacts with our biorhythms. What grounds us, biologically and spiritually, isn't just logic. It's harmony.

AI, on the other hand, does not exist in harmony. It exists in code. Fast, synthetic, weightless. But what if we used AI to assist, not override, these natural systems? What if, instead of trying to clone human decision-making, we used AI as a mirror to better understand our subconscious patterns? What if it became a compass that guided us toward deeper awareness rather than replacing our internal GPS?

AI should be a bridge to our higher selves, not a barrier between us and our instincts. To stay ethically awake in a world run by algorithms, try this:

- *Ask why before how.* Before using any AI tool, ask why it exists and who benefits.
- *Pause at the edge of automation.* When something feels "too easy," check if you're skipping an important emotional or moral process.
- *Journal your decisions.* Reflect on when and why you let AI take the lead.
- *Engage in values-based debates.* Invite discussions about right versus wrong, even when there's no clear answer.
- *Feel the friction.* Ethics often comes with discomfort. Don't rush past it.

Our subconscious patterns will always leak into the systems we create. That's human. But we have the power to reflect, refine, and realign.

AI is not the enemy. It's a tool. A powerful one. One that, if used with intention, could elevate our collective consciousness. But only if we stay present. Only if we remember that conscience cannot be coded. Let AI work with gravity, not against it. Let it assist your evolution, not mimic your mind. Let it help you think, but never forget how to feel.

Reclaiming the Brain: The Power of Conscious Resistance

So, what do we do? Go back to stone tools? No. We build with AI, but we live as humans. We let AI help but not replace. We choose to be confused sometimes. We let ourselves forget, wonder, fail, and feel awkward. But reclaiming the brain isn't just about intellect. It's about intuition.

That quiet nudge inside you, the one that says something feels off, even when the data looks perfect isn't just superstition. It's your internal radar. A complex, beautifully evolved fusion of subconscious signals, gut-brain communication, and emotional memory.

You've felt it before the moment you walked into a room and sensed tension before a word was spoken. Or the pause you took before saying yes to something that felt wrong. That's your intuition.

When we ignore it too long by over-relying on AI, by dismissing our instincts as old-fashioned, we risk turning down the volume on one of our most powerful tools. Reclaiming your brain means reactivating that deep sense of knowing without logic. It means trusting that your experiences, emotions, and quiet awareness often know things before your rational mind catches up.

Conscious resistance is about giving space for that voice to speak. But it's also about acting. Radical, real, grounded action. Here's what really works:

- *Take cold showers*. It's uncomfortable but that discomfort wakes your nervous system. It triggers a flood of dopamine, sharpens your focus, and makes you mentally tougher. People who do this daily report higher resilience and better emotional control.
- *Wake up early, before sunrise.* This isn't just about productivity. Early hours align with the body's natural circadian rhythm. The quiet rewires your brain for stillness, reflection, and intentional thought.
- *Do something hard, every day.* Challenge builds character. It doesn't have to be dramatic. Take the stairs. Learn something new. Apologize first. Read a difficult book. Growth lives where resistance starts.
- *Create before you consume*. Before checking your phone or email, write a page, stretch, meditate, or cook something. Send a message of intention to your brain: *I lead my day, not the world outside me.*

- *Adopt one new habit a month.* Your brain is plastic, it molds. You can train it to build discipline like a muscle. Try writing with your non-dominant hand. Or brushing your teeth in silence. These seem small, but they activate and rewire different brain centers.
- *Journal your thoughts in pen.* Handwriting slows your brain down just enough to make subconscious thoughts visible. When you write something emotional or intuitive, it integrates deeper.
- *Talk to someone with opposite beliefs.* Not to win—but to listen. This opens neural networks for empathy and flexible thinking.
- *Stand barefoot on the ground.* It's called grounding, and there's real science behind it. The Earth's negative ions can neutralize inflammation and stabilize your body's electrical systems. It's not spiritual fluff, its ancient wisdom backed by biophysics.
- *Fast not just from food, but from stimulation.* Silence, boredom, and solitude are healing. They're where imagination hides. They bring clarity the algorithm can't.

You don't need a full system overhaul. You just need to reclaim *one moment* a day. That one spark of raw humanity. That one pause before the app answers. That one time you say, "Let me think." That's enough to begin the reset.

Let your intuition breathe. Let your subconscious speak. Let your brain—not just the fast part, but the deep part guide you back to yourself. You can use AI. You can love it. But don't let it rob you of your depth. The brain isn't just a processor. It's a meaning-maker. And your meaning is yours to make.

About the Author

Sarah Choudhary is a global AI and data leader, researcher, speaker, and visionary known for weaving ethical consciousness into cutting-edge technology. With over 20 years of experience spanning AI, cloud systems, cognitive neuroscience, and innovation strategy, she stands at the intersection of logic and empathy.

Sarah is the founder of ICE Innovations and the creator of transformative platforms like iChef and ICE Ride, combining machine learning with real-world human needs. She holds a PhD in data science and is deeply involved in thought leadership around responsible AI, quantum computing, and neuro-AI research.

Her work has been featured across global forums, and she's known for her ability to decode complex systems into relatable, human-centered insights. Beyond tech, Sarah is a poet, painter, personal trainer/coach, and lifelong learner driven by curiosity and compassion.

Email: sarah@sarahchoudhary.com

Website: www.sarahchoudhary.com

CHAPTER 5

OVERCOMING AI'S ALIENATION AND ACHIEVING NEW HEIGHTS OF CREATIVITY AND FREEDOM

By Pedro Clark Leite
Project and Product Manager
Rio de Janeiro, Brazil

Freedom is what you do with what's been done to you.
—Jean-Paul Sartre

In the rapid rise of the AI revolution, the world stands at a precipice, peering into a future both daunting and mesmeric. Unlike the other recent trend buzzwords like "blockchain" or "quantum computing," AI is a reality that urges us to question not just how we will use it, but also how AI will reshape the very fabric of our existence.

As civilization rushes toward this brave new reality, we should consider a society where artificial intelligences govern not only the mundane tasks of daily life but also the complex decisions that shape economies and influence political spheres. Tech enthusiasts, like this author or you readers, could see the benefits of such a scenario where AI is employed to optimize economic strategies, predict market trends with unprecedented accuracy, and manage national or global economies to avoid recessions and ensure prosperity.

However, the other side must be taken into consideration: The entities that control AI technologies could potentially gain disproportionate power, leading to new forms of digital oligarchy or technocracy. This concentration of power might lead to the abuse of AI for personal or corporate gain, manipulating economic and political systems to favor certain groups over others. The disparity in access to technology, infrastructure, and expertise between developed and developing nations could widen existing inequalities, potentially exacerbating poverty and hindering development in several ways.

Looking into the microeconomics and our daily lives, there lurks another tangible threat—the specter of alienation, a concept Marx would recognize all too well. As AI takes over more of the tasks that have traditionally defined human labor, what becomes of the worker? Karl Marx's concept of alienation, which he primarily developed in the context of labor within capitalist systems, provides a critical framework for understanding the potential human consequences of an AI-dominated world. Marx argued that in capitalist societies, workers are alienated in four key ways: from the product of their labor, from the process of production, from their own species-being, and from other humans. The worker becomes a mere cog in the machine, estranged from their own humanity and the fruits of their labor.

Transposing this concept into the AI revolution, we see potential parallels that are both stark and unsettling. As AI systems become increasingly capable and autonomous, humans may find themselves alienated not just from traditional forms of labor, but from intellectual and creative endeavors as well. AI's ability to analyze data, generate insights, and even create art and write poetry could lead to a profound sense of displacement and redundancy. This alienation could be exacerbated by the rapid pace of technological change, making it difficult for individuals to find a stable sense of identity and purpose.

To view the issue solely through the lens of job displacement, a common practice by the big media and so-called influencers is to simplify a complex problem excessively. The integration of AI into our societal frameworks raises profound moral and ethical questions. It is here, in this profound existential reexamination, that the philosophical musings of another philosopher from the 19th century, Friedrich Nietzsche, find new life and urgency. The late philosopher challenged us to transcend the conventional moralities and truths of our time, urging us toward the creation of new values that are more robust, life-affirming, and reflective of our individual will to power.

In our transformed reality, Nietzsche's proclamation of the Übermensch takes on a new dimension. Here, the Übermensch is not a goal for a single individual alone but a challenge to rise as humans alongside our creations, to redefine our values in the face of an intelligence that might soon surpass our own. This AI, with its potential to transcend human limitations, compels us to question what it means to be human in the first place. Nietzsche encourages us to redefine values and meanings in the face of existential challenges. In the context of AI, this could mean harnessing the same technologies that threaten to alienate us to, instead, expand human capabilities and redefine labor and creativity. If AI can perform tasks more efficiently, humans might be freed to engage in more fulfilling, creative, and exploratory activities, thus overcoming alienation through a reconnection with their species-being and the essence of what it means to be human. The challenge, then, is not merely to integrate AI into our societies but to do so in a way that enhances human dignity rather than diminishing it.

Therefore, Nietzsche's concept of the will to power can be seen as an antidote to Marxian alienation. If humans can assert their will to power through the creation and control of AI, using these tools to enhance rather than diminish human agency, a new synthesis might be achieved. In this synthesis, AI becomes not a source of alienation, but a means of overcoming it, helping humans to achieve new heights of creativity and freedom. This would require a conscious effort to shape technological development in ways that enhance rather than replace human capabilities, fostering a society where technology serves humanity, and not the other way around.

Thus, integrating Nietzsche's philosophical insights with Marx's critique of alienation provides a dual lens through which to view the AI revolution: one that acknowledges the risks of dehumanization and estrangement, but also recognizes the potential for human enhancement and liberation. This nuanced perspective encourages a proactive approach to technology, advocating for a future where AI is integrated into a broader humanistic framework, promoting not just efficiency and productivity, but also meaning, purpose, and connectivity.

Achieving the independence that Nietzsche advocated allows us to also critically assess and potentially slow down decisions that may prove detrimental to humanity and the democratic process. Nietzsche's philosophy encourages the reevaluation and transformation of established values, urging individuals to assert their will to power—a concept that involves overcoming the external forces and internal limitations that constrain human potential. In the context of AI's expanding influence, this philosophical stance empowers us to question and contest the unchecked authority of technocrats who might prioritize efficiency and technological advancement over ethical considerations and human welfare.

Knowledge of the situation is paramount. By cultivating a deep understanding of AI's capabilities, limitations, and the ethical landscapes it navigates, society can equip itself to hold technocrats accountable and ensure that technological advancements serve the broader interests of humanity rather than a select few. This demands a robust public discourse on AI ethics, transparent reporting of AI developments, and inclusive policies that consider the diverse impacts of technology across different societal groups. Education and awareness-raising about AI should be prioritized to demystify the technology and empower individuals to participate meaningfully in decisions about its deployment.

In essence, the only way to prevent a world ruled by technocrats with the highest form of individual alienation—where decisions are made by a few, based on algorithmic calculations rather than democratic consensus—is through informed, engaged citizenship. Nietzsche's call for the revaluation of values and the assertion of individual agency resonates strongly in this context, urging a vigilant and proactive stance against potential overreach by powerful technological elites. This approach not only safeguards democratic principles but also ensures that technological progress aligns with the ethical and moral standards that define our humanity.

To thrive in this new era, we must become like the sailors of old who mastered the art of navigating treacherous, ever-shifting seas. We must learn to harness the capabilities of AI, directing its growth so that it serves not as a master but as a powerful tool for human betterment. It requires a bold reimagining of our social contracts, an embracing of change coupled with a steadfast commitment to the values that define our humanity—compassion, creativity, and a relentless quest for knowledge. Jean-Paul Sartre's existential assertion that "existence precedes essence" radically shifts the philosophical perspective on human nature, suggesting that individuals are not born with a predetermined purpose or essence, but rather, they must forge their own identities and meanings through their actions and choices. This foundational concept of existentialism emphasizes the profound freedom and responsibility each person holds in defining their existence.

As we stand facing this situation, looking into the future, we must decide whether we will be led into this new era or whether we will lead ourselves. Will we allow AI to shape our destiny, or will we seize the extraordinary tools at our disposal to carve out a future that reflects our highest aspirations?

In the end, thriving in the age of AI is not merely about survival or maintaining the status quo. It is about envisioning and creating a world where technology amplifies our human potential, not diminishes it. It is about ensuring that in this great symphony of digital and organic life, the human spirit continues to play the lead melody, soaring above the mechanistic harmonies in a powerful testament to our enduring quest for meaning and connection in an ever-evolving universe.

About the Author

Pedro Clark Leite is an awarded PMO senior project manager and customer service leader who has delivered projects for top tech companies such as Google, Apple, AT&T, TMO, and Verizon and worked as a PM consultant at Stanford and MIT. He has experience as a regional head, trainer, PO, and scrum master. Pedro is a certified PMP, Azure, AWS, and ITIL Agile coach with a talent for leading teams globally and defining processes and tools. He speaks four languages and has represented his companies in 15-plus countries.

Email: pedroclarkleite@gmail.com

LinkedIn: https://www.linkedin.com/in/pedroccleite/

CHAPTER 6

DOES AI POSSESS CREATIVITY AND ORIGINALITY?

By Jingying Gao, PhD
AI Scientist, AI & Robotics Lab, UNSW
Sydney, Australia

Creativity ignites where guided memory meets fruitful randomness—whether in a trillion-parameter model or the human mind.

—Large Language Model

The Birth of Artificial Intelligence

In the summer of 1956, a group of visionary young scientists gathered at Dartmouth

College in the United States. They carried with them a bold dream: to make machines think like humans. These scientists believed that computers could be taught to mimic human intelligence, solve complex problems, and perform tasks that previously required human effort. This gathering, later known as the historic "Dartmouth Conference," marked the formal birth of the field of artificial intelligence (AI) and gave rise to the term "artificial intelligence."

So, what exactly is artificial intelligence? Simply put, AI is the ability of machines to simulate human intelligence. It allows machines to "see," "hear," "read," "write," "sing," "paint," and even "think" like us. At the time, this goal seemed wildly ambitious. But with rapid technological progress, much of that early vision has already become reality today.

Modern AI can already "see" artwork and digits, identifying objects through computer-vision techniques that now help doctors detect tumors with roughly 95% accuracy. Beyond vision, AI has made rapid progress in language. It began by recognizing individual words as numerical tokens, then learned to model word relationships through embeddings—compact numerical vectors that encode each word's meaning and its similarity to other words—and later powered question-answering systems.

The transformer architecture, and with it the idea that "attention is all you need," pushed AI's grasp of language even further. Since 2022, led by ChatGPT, we have entered the era of generative AI which pushes AI's capabilities into the realm of creativity.

Creativity—one of humanity's unique capacities—is the ability to generate new ideas and to discover or invent new things. Cognitive-science scholar Margaret Boden, in *Creative Mind*, defines creativity as a flash of insight in which a novel idea emerges—one that no one has conceived before, shaping the thoughts of many and driving society forward. Works such as Picasso's paintings, the invention of the violin, masterpieces like Don Quixote and Journey to the West, the light bulb, and the steam engine are all products of such creativity.

Today's AI can already produce impressive works. It can draft a novel on the scale of *Three Kingdoms* for only a few dollars in token costs or create a detailed image from a single prompt. There's no denying that AI has reached a level of "master-class" creative ability. Yet many critics argue that AI lacks autonomous creativity. Instead, its "creativity" is rooted in statistical models that extrapolate from data to predict the next likely outcome—something they believe doesn't qualify as real creativity.

Does AI possess autonomous creativity at all? Will it lessen human creativity or even surpass it? To approach these questions we must first look at how the human brain creates, then examine how AI creates—for example, how it turns a text prompt into an image.

How Our Brain Creates

Creativity doesn't just rely on wild imagination. Recent neuroscience research by Roger Beaty, a post-doctoral fellow in psychology at Harvard University, used functional MRI (fMRI) to observe brain activity during creative tasks. The experiment invited 163 volunteers to enter an MRI scanner and to imagine unconventional uses for everyday objects. The findings show that creativity does not arise from a single brain area or from some lone "aha" moment; it is the product of several regions working together and involves three key neural networks.

First, the default mode network (DMN)—the source of imagination and association. The DMN activates when a person is relaxed, daydreaming, or "zoning out." This system helps us recall the past, envision the future, construct hypothetical scenarios, and engage in free association and fantasy. It acts as the brain's "engine of imagination," and many original creative ideas emerge from this system. A vital region within the DMN is the hippocampus. The hippocampus is central to human memory formation. Without it, not only would people suffer memory impairment, but they would also lose the ability to generate new ideas.

When the brain teems with imaginative notions, the second network, the salience network selects the most striking ideas from the default mode network and routes them to the executive control network for appraisal. The executive control network is the manager that filters ideas and applies logic. Although the default mode network supplies a steady stream of raw material, without a system of selection those thoughts would remain chaotic and impractical. Centered mainly in the prefrontal cortex, the executive control network evaluates emerging ideas and continually adjusts the direction of creative thinking.

Human creativity, therefore, moves from free association, where countless imaginative notions emerge, to logical appraisal, where workable ideas are refined. That dynamic has given rise to automobiles, inspired great paintings, and produced poetry and music that endure through time.

Decoding AI Creativity

If human creativity arises from imagination shaped by memory, logic, and evaluation, how does artificial intelligence create? To answer that, we can explore the underlying computational principles that power AI's creativity across different fields: visual art, writing, music, and scientific discovery.

Visual Creativity: How AI Generates Images and Videos

In the realm of image generation, one of the most groundbreaking innovations is the diffusion model, which draws inspiration from physics: when dye molecules are dropped into pure water, they disperse from areas of high density to low density—a process known as diffusion. Now imagine reversing this process: starting with water that is uniformly colored and trying to reconstruct the moment just after the dye first entered.

Inspired by the way dye molecules spread out, in the context of AI, a stable diffusion model works by gradually adding noise such as gaussian noise to an image through a series of steps, effectively destroying the image until it becomes pure noise. This is known as the forward diffusion process. Then, using a neural network—typically a U-Net architecture—the model learns to reverse this process, step by step, predicting and removing the noise to reconstruct a coherent image. This reverse diffusion allows the model to generate entirely new visuals from what starts as random noise.

Since then, AI models have learned to generate images from random noise, but the results are still essentially random. How can they synthesize images that correspond to a specific text prompt—for example, "a corgi playing a flame-throwing trumpet"? This is where CLIP (contrastive language-image pretraining) AI model comes in. CLIP is trained on massive datasets of image-text pairs. It learns to map both images and their associated text into a shared vector space. The model is taught to align image and text representations using contrastive learning: image encoders and text encoders convert their inputs into vectors, and the model is optimized to bring matching image-text pairs closer together while minimizing mismatched pairs.

When the human eye sees 24 images per second, it perceives them as smooth motion—a frame rate that underpins film and animation. Recent research pushes AI beyond single frames: a paper "one-minute video generation with test-time training" shows a model that produces a full, unedited minute-long story in one pass. Trained on Tom and Jerry clips, the system needs no stitching, post-processing, or manual tweaks, and—unless told otherwise—most viewers would struggle to spot that the resulting cartoon is entirely AI-generated. This breakthrough showcases a new frontier in machine creativity.

Creativity in Writing

ChatGPT has become a tool almost everyone reaches for. One person might ask it how to invest in properties near a city; another might use it to write a book that gets published or to build code for a website. Yet it can also hallucinate—sometimes confidently presenting fiction as fact.

Why can an AI suddenly draft poems, spin short stories, write code, or dispense surprisingly professional advice? The engine under the hood is the large language model (LLM). At its core, an LLM is trained on massive text datasets—from the internet, books, academic journals, code, and more—much like a professor who has read an entire library. During training, the model learns one basic task: predict the next token (a piece of a word) given all previous tokens. It does so with a transformer architecture, whose attention mechanism weighs how strongly each earlier token should influence the next prediction/ which uses a mechanism that calls attention to weigh the relevance of each word in the context of the sequence. Over billions of training steps, the model captures statistical patterns, syntax, semantics, and style.

With enough training data—billions of sentences—AI becomes remarkably good at continuing a thought, mimicking style, or even constructing coherent narratives. The process is statistical, yes, but it's also generative: It creates something new, not just by copying, but by recombining and reimagining patterns learned from language.

Creativity in Music

In music, platforms like Suno.ai can now generate full compositions from short prompts, complete with melody, harmony, lyrics, and instrumentation. The model predicts sound patterns that align with human musical structures and emotional tone. But this also raises new questions: Who owns the music? Does the model have rights? Does the user? In most current systems, the intellectual property (IP) is assigned to the user if they are a paid member, such as based on Suno.ai's policy, but the legal and ethical implications are still evolving. What is clear is that AI is not just performing; it's composing.

Is AI's Creativity Original?

In all these fields, the same question persists: Is AI's creativity original? AI does not dream, feel, or reflect the way humans do. Its creations are based on vast patterns in data, and some argue that this makes it derivative rather than truly original. Yet the outputs are new. They have never existed before. A painting no human has drawn, a melody no composer has written, a sentence no poet has imagined—created not by copying, but by learning the deep structure of creation itself.

So where do we draw the line between simulation and creativity? Between recombination and invention? That is the heart of the debate—and the frontier we are now crossing. Let's take AI-generated images as an example. How does AI create a painting or transform human language descriptions into generated images? For instance, how does AI create an image of "a cat playing the piano"?

Artificial intelligence uses algorithms to generate an image through a set of numbers. These numbers are not just ordinary numbers; they represent how AI perceives and understands the world—using numbers to represent every detail in an image, such as color, shape, and texture. In a sense, what AI does is to begin by randomly generating an image with plausible pixel features, then continuously check what the generated image looks like, making adjustments to eventually create an image people desire.

For example, if we ask AI to generate an image of "a cat playing the piano," it will randomly select a "seed" of numbers and then gradually adjust these numbers according to its learned rules, until it generates an image that looks like "a cat playing the piano." It's like a painter continually refining their sketch until they create the desired masterpiece. An image of "a cat playing the piano" is the result of AI optimizing and adjusting the "number space" based on the target we've set.

In reality, AI is highly statistical. Creating something original is quite simple—we just need to let the AI network randomly choose a sequence of random numbers, and those random sequences will often turn out to be surprising, creative, and original. However, whether the random numbers generate random images or random text, they don't carry much meaning for us. What we are truly interested in are things that are not only original but also "meaningful."

Take art creation as an example. Randomly drawing some pixel arrangements may be original, but they won't have much meaning. This is just like how the human brain can come up with wild, fantastical ideas, but without the logic filter from the control network, these ideas don't hold any real meaning.

If we want to explore the different possible values of embedding vectors in latent space and try to randomly change these numbers, what would happen in that space of numbers that might represent the meaning of things? If we change these numbers randomly, what would emerge? AI can, indeed, generate all kinds of new images of a cat playing the piano, but most of them may be completely unrecognizable, or they may not resemble the "cat" we had in mind at all. We will see various images, but most of them we cannot interpret. Perhaps some of these images might look interesting, but we cannot express their meaning with language. Therefore, AI has the ability to uncover unexpected, novel things, but this is not a difficult task.

When scientists conduct creative experiments, they often do so with a purposeful guide, just like how we discover new minerals or materials in nature. AI's creativity, in fact, is the ability to create something new and useful under the guidance of humans. This creativity is not just about generating random data; it's about combining human guidance with statistical randomness to produce something meaningful.

Why is this important? If we let AI operate freely, it can discover things humans have never thought of. While most of them may seem meaningless, occasionally there will be valuable discoveries, just like how we discover new minerals or materials in nature. This is akin to how we explore nature and discover new substances, realizing that these new materials have practical uses. This principle also applies to AI in language, such as in the creation of novels and music.

AI's creativity in novel writing begins when humans provide a few starting keywords. The AI algorithm then looks for the most likely next word by examining the position of those words in a large numerical dictionary, considering different dimensions. It continues to predict, but these predictions are not completely random. They are based on the semantic relationships the AI has learned from human language, generating meaningful text, novels, or scientific content.

As for AI-generated music, it can generate imaginative lyrics by randomly generating words and then create a coherent musical score to produce pleasant-sounding music. When people are amazed by AI-generated music, they should also acknowledge its originality, just like a music student who systematically learns music and then creates their own compositions. As for whether AI can perceive the emotions conveyed by the music it creates, we cannot yet definitively say yes or no.

Advanced, cutting-edge AI systems are now capable of self-reflection and revising their creative outputs. This creative process mirrors how the human brain moves from free imagination to logical evaluation—refining ideas into meaningful innovations and inventions that shape the world in lasting ways.

AI-driven creativity is already transforming—and will keep transforming—our lives in countless ways. Should we remember and celebrate these creative contributions, just as we honor the inventors, writers, and artists who have shaped human history?

References/Notes

1. Boden, M.A., 2004. *The creative mind: Myths and mechanisms*. Routledge.
2. Wolfram, S., 2023. Generative AI Space and the Mental Imagery of Alien Minds. *Retrieved May, 20*, p.2024.
3. Wikipedia. (2025) *History of artificial intelligence*. [online] Available at: https://en.wikipedia.org/wiki/History_of_artificial_intelligence
4. Beaty, R.E., Kenett, Y.N., Christensen, A.P., Rosenberg, M.D., Benedek, M., Chen, Q., Fink, A., Qiu, J., Kwapil, T.R., Kane, M.J. and Silvia, P.J., 2018. Robust prediction of individual creative ability from brain functional connectivity. *Proceedings of the National Academy of Sciences, 115*(5), pp.1087-1092.
5. Vaswani, A., Shazeer, N., Parmar, N., Uszkoreit, J., Jones, L., Gomez, A.N., Kaiser, Ł. and Polosukhin, I., 2017. Attention is all you need. *Advances in neural information processing systems, 30*.
6. Radford, A., Kim, J.W., Hallacy, C., Ramesh, A., Goh, G., Agarwal, S., Sastry, G., Askell, A., Mishkin, P., Clark, J.

and Krueger, G., 2021, July. Learning transferable visual models from natural language supervision. In *International conference on machine learning* (pp. 8748-8763). PmLR.

7. Rombach, R., Blattmann, A., Lorenz, D., Esser, P. and Ommer, B., 2022. High-resolution image synthesis with latent diffusion models. In *Proceedings of the IEEE/CVF conference on computer vision and pattern recognition* (pp. 10684-10695).
8. Suno. (n.d.) *What is Suno's copyright policy?* [online] Available at: https://help.suno.com/en/articles/2746945 [Accessed Sep 2024].

About the Author

Dr. Jingying Gao is a senior AI scientist with expertise in multimodal and generative AI, specializing in multimodal reasoning and explainable AI. She earned her PhD from the AI & Robotics Lab at UNSW and has published research in top-tier AI conferences including NeurIPS, IJCNN, and others. She has also served as a reviewer for leading AI conferences including NeurIPS, ECAI, and IJCAI, and co-organized a workshop at IJCNN. After completing her PhD, Dr. Gao took on a Senior Manager role in AI and Data Science, where she leads generative AI research and applied AI projects. She brings extensive experience across both academia and industry, with a strong background in AI Research & Development leadership. She is also passionate about cognitive robotics and previously co-founded a robotics startup focused on AI innovation. There, she led R&D teams to design, develop, and deliver two family service robots, Yiling and Yiyi.

Email: jingying.gao@unsw.edu.au

LinkedIn: https://www.linkedin.com/in/gaojingying/

CHAPTER 7

RETHINKING WORKER SAFETY WITH AI

By Vivek Gnanavelu
Founder, Intrapreneur, AI Expert
Plano, Texas

The real enemy of safety is not non-compliance but non-thinking.
—Rob Long

Worker safety is a cornerstone of any responsible and sustainable organization. In industries like construction, manufacturing, mining, oil and gas, and logistics, the risks faced by frontline workers are significant. Traditional safety practices, while foundational, are often reactive, responding to incidents after they happen. The integration of artificial intelligence (AI) in workplace safety represents a revolutionary shift toward proactive, data-driven safety management. This chapter explores the importance of using AI for worker safety, the ways in which it is transforming industries, and the profound impact it can have on lives, costs, and organizational culture.

Worker safety is a universal priority, and regulations are tailored to protect employees in every industry. In the United States, the Occupational Safety and Health Administration (OSHA) sets and enforces standards to ensure safe and healthy working conditions across a wide range of sectors, including construction, manufacturing, healthcare, transportation, agriculture, and more. Employers are legally required to provide a workplace free from recognized hazards, regardless of the industry, and must comply with all applicable safety standards.

Scope of the Problem

According to the International Labor Organization (ILO), 2.3 million people die annually due to work-related accidents or diseases. Every year, there are approximately 340 million occupational accidents. In the United States, the Occupational Safety and Health Administration (OSHA) reports that over 5,000 workers died on the job in 2022—an average of nearly 100 per week. Here are some serious, up-to-date numbers on worker safety in the United States:

- *Fatal work injuries*: In 2023, there were 5,283 fatal work injuries, which equates to 3.5 fatalities per 100,000 full-time equivalent (FTE) workers.
- *Nonfatal workplace injuries and illnesses*: Private industry employers reported 2.6 million nonfatal workplace injuries and illnesses in 2023, an 8.4% decrease from 2022.
- *Incidence rates*: The total recordable case (TRC) rate in private industry was 2.4 cases per 100 FTE workers in 2023, the lowest since 2003.
- *Days away from work*: There were 946,500 nonfatal injuries and illnesses involving days away from work in 2023, representing 62% of all cases involving days away, job restriction, or transfer.
- *Worker deaths over time*: Worker deaths have dropped from about 38 per day in 1970 to 15 per day in 2023, but

preventable deaths still occur at a rate of three per 100,000 workers.

- *Occupational diseases*: An estimated 120,000 workers die each year from occupational diseases, a figure much higher than deaths from acute injuries.
- *Disparities*: Black and Latino workers face higher fatality rates—black workers at 4.2 and Latino workers at 4.6 per 100,000, both above the national average.
- *Older workers*: Workers aged 65 and older have a fatality rate of 8.8 per 100,000, 2.4 times higher than other workers.
- *Penalties*: The median federal OSHA penalty for a worker fatality in 2023 was $14,063.

These statistics highlight the ongoing challenges in worker safety across all industries in the U.S.

High-risk industries:

- *Construction*: falls from height, machinery-related accidents
- *Manufacturing*: equipment injuries, exposure to toxic substances
- *Mining*: cave-ins, gas explosions, respiratory illnesses
- *Logistics*: vehicle collisions, fatigue-related accidents
- *Oil and gas*: explosions, chemical exposure, high-pressure equipment failures

The scale of risk, both in human and financial terms, makes worker safety a pressing concern for organizations and governments alike.

Traditional Safety Practices and Their Limitations

Conventional approaches to workplace safety include:

- Training sessions
- Compliance checklists
- Incident reports and manual inspections
- Safety audits
- Personal protective equipment (PPE) checks

While effective to a degree, these methods are:

- Reactive rather than preventive
- Manually intensive, leading to delays in hazard identification
- Error-prone, especially when relying on self-reporting
- Static, failing to adapt to real-time changes on the ground

This is where AI can make a transformational difference.

How AI Enhances Worker Safety

Real-Time Hazard Detection

AI-powered computer vision systems can analyze video feeds from cameras installed on job sites, warehouses, or production floors. These systems can identify:

- Unsafe behaviors (e.g., not wearing PPE)
- Dangerous proximity to machines
- Slips, trips, and falls
- Violations of safety zones

Example: A worker enters a restricted zone without a helmet. The AI system detects the violation and sends an alert to the supervisor within seconds, preventing a potential injury.

Predictive Analytics

AI can analyze historical data—accident logs, weather data, equipment performance—to predict high-risk scenarios before they happen. AI can:

- Predict fatigue-related incidents using work schedules and biometric data
- Forecast machinery failure based on usage patterns
- Identify trends that signal increased accident likelihood

This predictive capability helps in reallocating resources, modifying schedules, or implementing extra safety protocols.

Wearable AI Devices

Smart wearables equipped with sensors and edge AI can monitor:

- Worker location (via GPS)
- Heart rate and body temperature
- Posture and movement patterns
- Exposure to hazardous gases

If the AI detects abnormal readings (e.g., signs of heat exhaustion or a sudden fall), it can trigger emergency protocols, including calling for medical help or shutting down nearby machinery.

Use of Large Language Model (LLM) for Incident Reporting and Audits

LLM can streamline the traditionally tedious process of safety reporting. It does this by:

- Converting voice notes to structured incident reports
- Automatically extracting key safety issues from logs

- Analyzing worker feedback to spot emerging safety concerns

This enables faster documentation, better compliance, and richer safety insights.

Benefits of Using AI for Worker Safety

Saving Lives

The most significant impact of AI in worker safety is its potential to prevent injuries and fatalities. Real-time alerts, predictive models, and automated responses enable action before it's too late, transforming safety from a compliance task into a life-saving mission.

Reducing Costs

Workplace injuries are expensive. In the US alone, the total cost of work injuries in 2022 was estimated at $167 billion, including:

- $48 billion in wage and productivity losses
- $36 billion in medical expenses
- $61 billion in administrative expenses

AI reduces these costs by preventing accidents, improving recovery times, and automating compliance.

Improved Regulatory Compliance

AI systems can ensure that companies meet OSHA and other regulatory body requirements by:

- Monitoring PPE usage
- Ensuring equipment maintenance

- Automating documentation for audits

This minimizes the risk of fines, shutdowns, and lawsuits.

Better Safety Culture

Implementing AI in safety programs signals a commitment to employee well-being, which can:

- Improve morale and job satisfaction
- Reduce turnover
- Attract top talent
- Strengthen trust between workers and management

When workers see that technology is used to protect them rather than monitor them, it reinforces a positive safety culture.

Some of the Areas Where AI Can Help Across Industries

Construction

- Drones with AI monitor construction sites for unsafe scaffolding, missing guardrails, or structural hazards.
- Computer vision systems detect when workers fail to wear hard hats or harnesses.

Manufacturing

- Machine-learning models predict equipment failure before breakdowns.
- Wearables detect ergonomic risks by analyzing repetitive motions.

Logistics and Warehousing

- AI-powered robots help reduce worker strain from heavy lifting.
- Computer vision/AI can be used to identify heavy lifting patterns to avoid ergonomics related injuries.

Oil and Gas

- AI detects gas leaks via drones or fixed sensors.
- Predictive maintenance avoids catastrophic equipment failure.

Mining

- Underground AI sensors monitor air quality and structural integrity.
- Smart helmets provide real-time alerts about dangerous gases or cave-in risks.

Ethical Considerations and Challenges for Adoption

While the advantages are clear, deploying AI for worker safety also comes with challenges.

Privacy Concerns

Surveillance-based AI systems may be perceived as invasive. Employers must:

- Be transparent about data use.
- Ensure data is anonymized.

- Avoid using safety data for disciplinary action unless justified.

Bias in AI Models

AI trained on limited datasets might fail to generalize across different work environments or demographics. This could lead to:

- False positives or missed hazards
- Unfair targeting of certain groups

Ethical AI practices require diverse training data, continuous monitoring, and fairness audits. This can be achieved through careful curation of relevant safety data from various parts of the world and industries.

Adoption Barriers

The following are barriers to adopting AI for worker safety:

- The cost of implementation may be high for small businesses.
- There might be resistance to change from workers unfamiliar with new tech, particularly workers who are used to traditionally operating with paperwork.
- There could be extensive training needs for both workers and safety managers.

These issues can be addressed through clear ROI communication, worker involvement, and phased rollouts.

The Future of AI in Worker Safety

As AI technology evolves, we can expect:

- Emotion AI that detects signs of stress or burnout
- Collaborative robots (cobots) that work safely alongside humans
- Augmented reality (AR) helmets with real-time hazard overlays
- Edge AI that processes data directly on devices, improving response times

Eventually, AI will be embedded in all layers of worker safety—from prevention and detection to response and recovery.

Summary

AI is not a silver bullet, but it is a game changer in protecting workers from harm. It augments traditional safety measures with speed, precision, and foresight that no human can match at scale. From detecting unsafe conditions in real time to predicting accidents before they occur, AI empowers organizations to go beyond compliance and truly prioritize human lives.

In a world where technology increasingly shapes how we live and work, integrating AI into safety programs is not just an innovation—it's a responsibility. It reflects a future where every worker can go home safe, every day.

About the Author

Vivek Gnanavelu is an intrapreneur and a data scientist with a strong engineering background. His expertise in AI and passion for worker safety created impactful solutions at scale within large enterprises like Ericsson. He is passionate about worker safety and strongly believes in saving human lives using technology.

Email: vivekgnanavelu@gmail.com

LinkedIn: https://www.linkedin.com/in/vivekgnanavelu

CHAPTER 8

BEYOND THE AI HYPE: CRAFTING A STRATEGIC ROADMAP FOR REAL BUSINESS IMPACT

By Yuri Gubin
Chief Innovation Officer, Advisor, Board Member
New York, New York

When you're finished changing, you're finished.
—Benjamin Franklin

The allure of artificial intelligence (AI) is undeniable. Headlines trumpet its transformative power, promising unprecedented efficiency, groundbreaking innovation, and unparalleled competitive advantages. Yet, navigating this landscape of hype and potential requires more than just enthusiasm; it demands a strategic roadmap grounded in a deep understanding of your organization's goals and a realistic assessment of AI's capabilities. This chapter will guide you through the essential stages of crafting such a roadmap, moving beyond the abstract promises to tangible business impact.

Part 1: Setting the Stage—Looking at the Landscape

Before diving into specific technologies or use cases, it's crucial to establish the fundamental purpose of AI within your organization. This initial stage is about strategic thinking—defining how AI will contribute to your core value proposition. What are the big things that define the map that you need to keep in mind? What is the landscape of the transformation? Forget the buzzwords for a moment and consider the following foundational elements:

The Creation of Value

What specific value will AI bring to your business? Will it power new products or features that offer unique customer benefits? Can it solve existing customer problems with greater speed and efficiency? Will it unlock entirely new capabilities that differentiate you in the market? This declaration acts as your North Star, guiding all subsequent AI initiatives. For instance, a healthcare company might declare AI's value in enabling faster and more accurate diagnoses, while an e-commerce platform could focus on AI-powered personalized recommendations to enhance customer experience and drive sales.

Value Capturing

Beyond creating new value, AI offers significant potential in optimizing existing operations. Where can AI automate repetitive tasks, freeing up valuable human capital for more strategic endeavors? Think about streamlining workflows, automating data entry, or enhancing communication. Even seemingly simple applications like automated transcription, meeting minutes, and content summarization can yield substantial productivity gains. Moreover, AI-powered tools can boost productivity across various functions, from assisting engineers with coding to empowering marketing teams with content creation and accelerating the development of internal strategies and policies.

Strategic Thinking vs. Agility of the Company

Declaring a strategy is only the first step. The dynamic nature of the business environment—evolving markets, technological advancements, competitive pressures, and regulatory changes—necessitates continuous strategic thinking and adaptation. AI itself can be a powerful catalyst for this evolution. By experimenting with AI, organizations can gain insights into their own agility and capacity for change. How effectively can your current structure and culture embrace new technologies? Do you have the power to drive the change? How quickly can you iterate? How much would the bureaucracy slow the innovation? These are among the questions you need to find answers to when thinking about AI adoption. Consider AI implementation not just as a technological endeavor but as a litmus test for your organization's ability to adapt and modernize in the face of future disruptions. This strategic thinking, this constant re-evaluation, is paramount to ensuring that your AI initiatives remain aligned with your evolving business goals.

Part 2: Identifying Big Landmarks on the Map

Once the strategic intent and landscape are clear, the next step involves identifying and addressing the big landmarks vital for the roadmap planning. These are the foundational elements required for successful AI implementation. Many promising proofs of concept (POCs) stumble not due to technological limitations, but because the underlying organizational infrastructure and support systems are inadequate. The roadmap needs to consider visiting these areas, developing capabilities, and making changes. These landmarks are crucial for translating AI potential into tangible results, such as the following:

Executive Support (The AI Committee)

Strong leadership buy-in is paramount. An AI committee, comprising senior management and executives, provides the necessary oversight, resources, and strategic direction for AI initiatives. Without this high-level support, AI projects risk being marginalized, underfunded, and lacking the organizational leverage needed for widespread adoption.

AI Policy and Vision

A clearly defined AI policy establishes the guiding principles for how the organization will work with AI. This includes the vision for AI adoption, acceptable use guidelines, data sharing protocols, and ethical considerations. As AI penetrates various aspects of the business, a formal policy ensures consistency, mitigates risks, and fosters responsible innovation.

Legal and Compliance

Navigating the legal and regulatory landscape surrounding AI is critical. Without adequate support IT and business will be operating blindly and mistakes can be expensive. This requires educating the legal and compliance departments on AI capabilities and implications, defining the boundaries of what AI can and cannot do, and establishing guidelines for data usage, model deployment, and decision-making. Proactive engagement ensures compliance and mitigates potential legal risks.

Enterprise Risk Management (ERM)

A robust ERM framework is essential for identifying, assessing, and mitigating risks associated with AI implementation. This proactive approach allows organizations to anticipate potential pitfalls, make informed decisions, and develop strategies to minimize negative impacts. Without a comprehensive ERM strategy, AI POCs are unlikely to scale into production-ready solutions with widespread adoption.

Information Security (InfoSec)

As AI systems handle increasingly sensitive data and become integrated into critical workflows, information security raises to be the top concern. Organizations need to equip their InfoSec teams with the knowledge and tools to understand AI-specific threats and vulnerabilities, enabling them to build and maintain secure AI applications and ensure the safe use of AI productivity tools.

Observability

Understanding the performance, cost, and impact of AI initiatives is a requirement for making informed decisions and demonstrating ROI. Observability encompasses both technical metrics (e.g., model accuracy, latency) and business-oriented KPIs (e.g., user engagement, cost savings). Continuously developing the observability component ensures that AI investments are delivering tangible value.

Data Management and Maturity

The quality, accessibility, and governance of data are fundamental to successful AI. This includes not only traditional operational data but also previously untapped sources like customer interaction recordings, product images, and documentation. Data needs to be structured and prepared for AI consumption, and organizations must understand the ethical and legal implications of using different data types.

Technical and Non-Technical Skills

It is impossible to build, deploy, and utilize AI effectively without skilled workforce and talent. This requires a comprehensive upskilling and training strategy that addresses the diverse needs of different roles within the organization. Establishing a culture of continuous learning and research helps staying abreast of the rapidly evolving AI landscape and identifying new opportunities.

AI Infrastructure

A robust and scalable AI infrastructure, encompassing the necessary hardware, software, cloud services, and integration capabilities, is where AI development and deployment is actually happening. This infrastructure needs to be strategically aligned with the organization's overall IT architecture to ensure seamless integration, efficient resource utilization, and robust security.

By thoughtfully visiting these landmarks, organizations can create a solid roadmap for their AI journey, significantly increasing the likelihood of successful implementation and realizing tangible business value.

Part 3: Setting the Compass—Defining Your Starting Point

With a clear landscape, strategic intent, and the landmarks identified, the next step is determining where to begin your AI journey. The sheer breadth of potential AI applications can be overwhelming, making a deliberate and informed starting point essential. A self-assessment, often conducted with the help of an experienced partner, can provide valuable insights into your organization's current readiness and identify promising initial opportunities.

The Value of External Perspective

Engaging with a partner who has already navigated the AI landscape can bring a fresh perspective, unbiased insights, and valuable experience to the assessment process. They can help identify blind spots and provide a more objective evaluation of your organization's strengths and weaknesses in relation to AI adoption.

Three Essential First Steps

Based on practical experience, the following three key actions provide a strong starting point for your AI roadmap:

1. *Appoint an AI champion.*

Designate a specific individual or a small, focused team led by an AI champion. This person will be the central point of contact for all AI-related initiatives, responsible for driving the strategy forward, communicating progress, and advocating for resources. This role, potentially a temporary assignment for a leader like the CTO or head of engineering, provides clear ownership and accountability.

2. *Establish an AI committee.*

Formalize an AI committee comprising representatives from key departments and senior leadership. This committee provides the necessary leverage, resources, and transparency for AI initiatives. It facilitates cross-functional collaboration, ensures alignment with business goals, and fosters a sense of urgency and executive buy-in. As an example, include a CEO in order to facilitate AI inclusion into company strategic discussions. The team should have the power, leverage, and influence.

3. *Create an AI center of excellence (CoE).*

Establish a forum or AI CoE, not necessarily involving the physical relocation of personnel, but rather a virtual or collaborative group that allows employees working on AI projects across different departments to connect, share knowledge, discuss challenges, and disseminate best practices. This fosters a culture of learning, accelerates the accumulation of organizational AI expertise (including prompts, architectures, and shared resources), and promotes cross-pollination of ideas. These three initial steps lay the groundwork for a structured and collaborative approach to AI adoption, setting the stage for more focused exploration and implementation.

Part 4: Navigating the Terrain—Embracing Change Management

Implementing AI is not merely a technological upgrade; it's a significant organizational change that requires careful management. Convincing stakeholders and employees to embrace AI is often a greater challenge than the technical implementation itself. Understanding the dynamics of change adoption is crucial for a successful AI roadmap.

The Learning and Adoption Curves

When introducing new technologies like AI, it's reasonable to anticipate the initial dip in productivity as employees learn new tools and processes. This "learning curve" is a natural part of the adoption process. Simultaneously, the "adoption curve" highlights the varying rates at which individuals embrace new technologies. You'll encounter early adopters, a chasm to bridge before reaching the early majority, a late majority who follow the trend, and inevitably, some who resist change.

Addressing Pain Points for Adoption

To facilitate AI adoption, focus on addressing specific pain points that employees experience in their daily work. Instead of presenting AI as a generic solution, demonstrate its value in solving concrete problems. When individuals experience tangible benefits—increased efficiency, reduced workload, improved accuracy—they are more likely to build confidence in AI and embrace its use.

By acknowledging and proactively managing the learning and adoption curves, and by focusing on solving real problems for employees, organizations can foster a more receptive environment for AI adoption and accelerate its integration into daily workflows.

Part 5: Charting the Course—Prioritization and Strategic Execution

With the initial groundwork laid and an understanding of the change management involved, the final stage involves identifying and prioritizing potential AI use cases and AI adoption projects, and strategically planning their execution to maximize impact and minimize risk.

Transparent Prioritization

When faced with multiple potential AI applications and projects, a transparent prioritization process involving all relevant stakeholders is crucial. Clearly communicate the criteria used for decision-making, ensuring that everyone understands why certain projects are being pursued first.

Considering ROI and Execution Risk

While the potential return on investment (ROI) of each use case is a key factor, it's equally important to consider the risk of execution. High-risk projects, especially early on, can lead to setbacks and erode confidence in AI. Prioritize "quick wins"—projects with a relatively high ROI and low execution risk—to build the momentum, demonstrate value, and foster trust in AI's capabilities.

Managing Complexity

Especially in the initial phases, focus on AI applications with lower implementation complexity. Success in these areas builds confidence and provides valuable learning that can be applied to more complex projects later. A "fail fast" approach within these simpler use cases allows for rapid iteration and course correction.

The Risk of Inaction

While carefully considering risks is important, also evaluate the potential risks of *not* pursuing certain AI applications. Are there opportunities being missed? Could inaction lead to a competitive disadvantage? This perspective can help identify strategically important use cases that might otherwise be overlooked.

Strategic Interconnectivity—The Bowling Alley Effect

When planning the execution of your AI roadmap, think strategically about how different AI initiatives can build upon each other. Aim for a "bowling alley" effect, where early successes in one area add up and make subsequent implementations easier, faster, and cheaper. For example, adopting AI-powered productivity tools for software engineers might streamline the development of AI-driven features for your product. Prioritize use cases that have a cascading effect, unlocking value across multiple departments or product lines. By employing a transparent, risk-aware, and strategically interconnected approach to prioritizing and executing AI initiatives, organizations can chart a course that leads to sustainable AI adoption and significant business transformation.

In conclusion, crafting a strategic AI roadmap is a journey that extends far beyond the technological aspects. It requires a clear understanding of your organization's strategic goals, a meticulous assessment of foundational requirements, a thoughtful approach to change management, and a pragmatic strategy for prioritization and execution. By embracing these principles, organizations can move beyond the hype and harness the true power of AI to drive real and lasting business impact.

About the Author

Yuri Gubin is a technology leader, practitioner, and advisor who empowers organizations to innovate, manage change, and resolve intricate technology challenges to revitalize their operations. His expertise spans enterprise architecture, cloud technologies, AI, and leading technology teams. Driven by a passion for architecture, he excels at identifying solutions for both technical and organizational issues, ranging from solution architecture to IT governance frameworks. He is a Stanford LEAD alumnus and NACD.DC director.

Email: yuri.a.gubin@gmail.com

CHAPTER 9

NAVIGATING THE AI DISRUPTION, FROM ENTERPRISES TO INDIVIDUALS

By Julien Guille
Head of Digital Acceleration
Tokyo, Japan

The best way to predict the future is to create it.
—Peter Drucker

OpenAI shook up the world with ChatGPT's release in late 2022, reaching millions of users in just two months—the fastest adoption of any technology. Behind its chat interface was GPT, a type of large language model (LLM) powered by a new deep-learning architecture. As the first of its kind, it paved the way for a wave of new models, chatbots, and tools, launching the generative AI era.

A portion of my role involves monitoring the digital innovation ecosystem. Over the past three years, barely a week has passed without a major breakthrough. Unlike recent disruptions (cloud, blockchain, etc.), AI has impacted both end users and businesses at an unprecedented speed. While we're clearly in a bubble, separating hype from reality is difficult—but one thing is certain: No one can afford to stand still. Enterprises must strategize, plan, and adapt, while individuals must upgrade their skills. While I can't disclose proprietary details, the insights I discuss in this chapter come from real examples I've worked on and seen in action. I'll start broad, then dive into how you—as a leader, professional, or individual—can navigate and thrive in this AI era.

Part 1: Understanding the New AI Wave

Artificial intelligence (AI) refers to the capability of computational systems to perform tasks typically associated with human intelligence, such as learning, reasoning, problem-solving, perception, and decision-making.

—Wikipedia

History of AI

Artificial intelligence as a field started in the 1950s with semantic reasoning and rule-based systems—this was also when the famous Turing test was introduced to assess machines' intelligence. After cycles of "AI winters" and "summers," the 2000s brought machine learning, shifting focus from programmed rules to data-driven learning. But limited computing power kept real-world applications out of reach. In 2012, deep learning emerged, enabled by GPUs (with their parallel computing capabilities), along with a vast amount of training data that the world had accumulated. This breakthrough paved the way for the latest revolution—generative AI—primarily based on deep learning and made possible by a key architectural innovation: the transformer.

Generative AI (GenAI) in a Nutshell

Fundamentally, GenAI creates content—text, images, videos, or music—by predicting the next element based on patterns from vast datasets. It breaks input into small units (e.g., tokens) and generates outputs probabilistically. Since it predicts rather than follows strict rules, results can be inaccurate or biased.

To guide AI responses and improve relevance, several techniques exist—each growing in complexity and typically handled by different roles. "Prompting" lets users shape outputs with clear instructions. IT leads contextualization—currently done with a RAG (retrieval augmented generation)—by adding enterprise domain data sources (e.g., company documents) and expanding the model's knowledge without altering it. "Fine-tuning" adjusts internal parameters using specialized datasets, requiring data science expertise. These methods refine predictions but don't eliminate uncertainty—so it is critical for users to always question and verify the output.

GenAI Latest Advancements: Beyond Generation

"Because it talks, it can understand, reason, decide, and act." This concise statement captures the recent breakthrough in generative AI—which truly stunned me once I understood what it can now do and how it does it (see chart below!). It now goes beyond content creation, extending into tasks that require reasoning and autonomous decision-making. Reflecting on this, you might wonder: Without language, would humanity have reached its current level of development? Perhaps a topic for another discussion ...

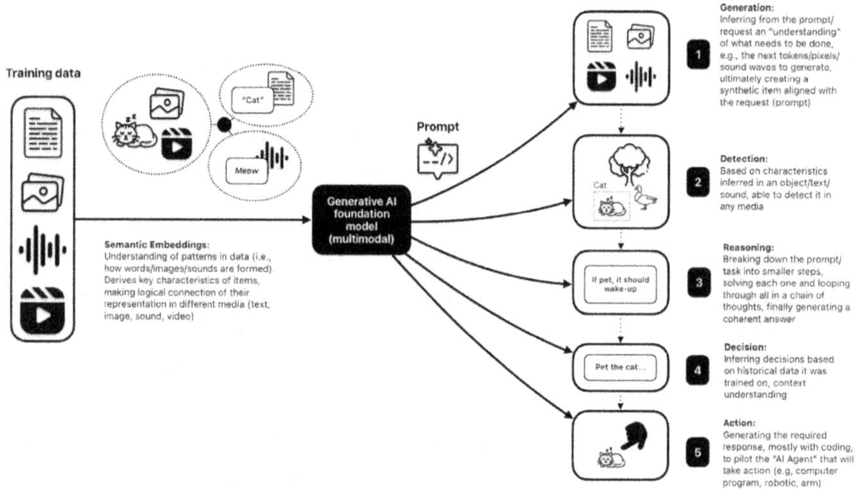

FIG. 1: FROM CONTENT CREATION TO COGNITION: HOW GENERATIVE MODELS ARE EVOLVING INTO AUTONOMOUS REASONING SYSTEMS

These advancements are already taking shape in two key areas:

- *In the virtual realm*: "AI agents" capable of autonomous decision-making are on the rise (for example in coding with AI coding agents), and emerging "agentic AI" systems aim to orchestrate multiple agents to accomplish complex, multi-step tasks. As of July 2025, this field is expanding rapidly, though still early in maturity. Imagine an AI super assistant that plans and books your vacation—handling transport, accommodation, and activities based on your preferences and budget.
- *In the physical world*: Vision language models (VLMs) and vision language action (VLA) models, such as Google's Gemini Robotics, are enabling robots to perform an increasingly wide range of tasks. While still under development and not yet perfect, these models are already

laying the groundwork for applications like autonomous driving.

While the full impact may not yet be evident, these developments show that GenAI—currently augmenting traditional models—could soon replace many of them, bringing us closer to scenarios once limited to science fiction.

Are Things Accelerating?

Indeed, AI product development has accelerated exponentially, driven by the flexibility of generative models that can handle multiple modalities (e.g., text, images, audio), require less labeled data for training, and are accessible even to non-experts through intuitive chat interfaces.

Taking a step back, this acceleration is happening across the entire tech industry—and it has surely fueled the rapid rise of AI development. Whether it's due to more powerful chips, technologies that enable massive data handling, or the simplification of architectures (e.g., APIs, containers, serverless cloud)—these building blocks are now more seamlessly integrated, drastically accelerating the development lifecycle of digital products, including those powered by AI.

Will It Last?

After two decades in tech, I see clear parallels between this acceleration and past hypes—like blockchain. But this AI disruption feels more foundational, closer in scale to the PC boom, internet wave, or smartphone era—transformations that reshaped society. Like those, I believe the AI hype will slow before settling into a lasting plateau. My view is shaped by:

- Gartner's 2024 Hype Cycle, which places generative AI on the "Trough of Disillusionment."

- Stanford's 2025 AI Index, showing that in 2024, fewer than 10% of companies reported AI impacting more than 10% of costs or revenue.
- AI will—and should—fade into the background as it integrates into everyday tools, with users simply expecting things to "just work"—the "AI effect."

Part 2: Fundamentals for Enterprise and Personal Readiness

As technology advances, distinguishing between transient trends and enduring principles is crucial for both enterprises and individuals.

Transformations to Adapt To

Enterprises are rushing to integrate generative AI to create new products and hyper-personalized services. This shift accelerates value creation, planning, and time-to-market. McKinsey's 2023 report "The Organization of the Future: Enabled by GenAI, Driven by People" estimated that up to 50% of business activities could be automated sooner than expected. While that forecast now invites nuance, the takeaway is clear: Data and AI are increasingly critical, demanding strong governance and continuous upskilling.

Leadership roles are evolving in this AI era. Executives are expected to be tech-savvy visionaries and embed digital tools into strategy and operations. As analytics increasingly shape decisions, leaders must know how to use them effectively—while remaining accountable for AI-driven outcomes and risks.

For individuals, AI literacy is becoming essential across most professions. The WEF's "Future of Jobs Report 2025" highlights rapid shifts in job roles, with skill mixes projected to change 65% by 2030. With AI copilots reshaping workflows, ongoing upskilling—even self-initiated—is now a must.

Foundations to Bank On

Despite technological shifts, core business principles remain firm. Companies still need clear strategies, customer focus, and strong governance—AI is a means, not the goal. Competitive edge still comes from human capital and innovation, alongside technology. Trust, brand reputation, and compliance remain essential, now including AI usage.

Core leadership traits—vision, judgment, and accountability—still matter. Executives must set ethical standards and culture; AI can't replace human qualities like inspiration or final decision-making. Long-term thinking and accountability toward stakeholders remain essential, even in times of tech disruption.

At the individual level, soft skills and ethics continue to define professional excellence. Communication, empathy, problem-solving, and integrity remain in high demand. Domain expertise, human creativity, and critical thinking are vital; AI tools assist but do not supplant the need for human insight and oversight.

In Summary

The World Economic Forum projects that by 2030, 92 million jobs will be displaced, while 170 million new roles will be created, resulting in a net employment increase of 7% jobs globally. This significant market shift presents opportunities for those prepared to adapt and thrive in the evolving landscape.

Part 3: Thriving Amid Disruption: Enterprise-Level and Leadership Priorities

In today's fast-changing world, companies must adapt to strong external forces. This section isn't a full playbook, but it proposes a high-level structured approach—preceded by a few key contextual points—that executives should keep in mind.

Setting the AI Agenda in Context

As AI gains traction, executives must look beyond the hype and adopt a broader, grounded perspective—one that considers market context, broader strategy, and that is cognizant of limitations:

- *Context*: In the ongoing digital revolution, companies are expected to serve increasingly tech-savvy customers.
- *Strategy*: True AI value emerges only when embedded within a broader digital strategy that, once executed, drives product and service enhancement as well as streamlined operations. Key transformation axes—often referenced in leading consulting frameworks—include strategy and roadmap, people and culture, customer experience, products and services, operations, and technology (including data and AI).
- *Limitations*: Several factors must be considered—slow workforce adaptation to new tools and work methods; scaling challenges that fall short of expectations set during accelerated prototyping; the need for clean, well-governed data, which is costly and time-consuming to manage; and finally, a GenAI cost and carbon footprint up to ten times higher than typical web usage.

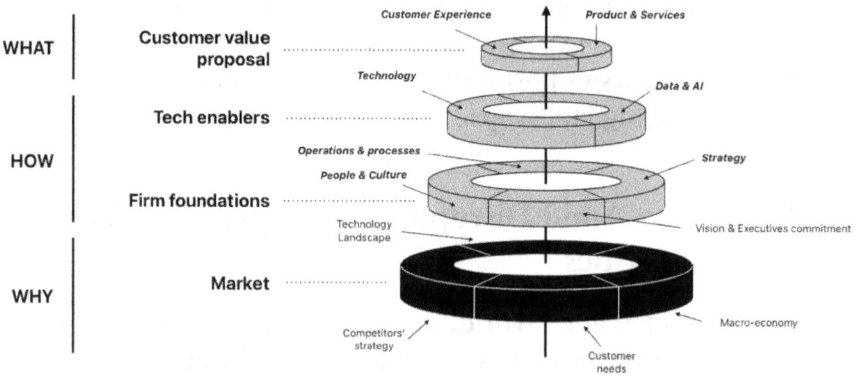

FIG. 2: ADAPTATION OF THE SDG WEDDING CAKE MODEL TO DIGITAL TRANSFORMATION. TECHNOLOGY LAYER CANNOT BE TAKEN IN ISOLATION TO PIVOT A COMPANY.

Defining the Strategic Direction

According to the McKinsey's June 2023 article, "How to Implement an AI and Digital Transformation," a successful AI integration starts with a business-led digital roadmap. Even in an era that prizes agility, a clear strategy remains essential—it sets the North star (the "why," "how," and "what") to deliver customer value. Regardless of the industry, C-suite alignment on a unified vision is key, with digital serving as an enabler, not the ultimate goal. Strategy frameworks (e.g., VMOST, Strategy House) typically stress two main phases:

- *Strategic intent:* Purpose → Objectives → Strategy
- *Execution:* Tactics and Plan → KPIs

AI fits naturally within the "Tactics and Plan" phase. Yet, today's GenAI hype often drives companies to rush into adoption—prioritizing technology over true business needs. Fragmented AI initiatives across departments can lead to increased risks, redundant efforts, and inflated costs. Executives should, therefore, quickly establish governance guardrails, form cross-functional GenAI teams, and clearly define—together—the purpose, approach, and goals for AI usage.

Adapting Enterprise Capabilities

With a clear strategic vision established, the next imperative is to refine the core capabilities—people, processes, and technology—that will transform that vision into operational excellence.

- *Talents*: Leadership must innovate in partnership with HR by creating market-aligned roles that are well integrated into the organization; defining a digital skills matrix and clear career paths; increasing insourced talent (targeting 70% to 80%); improving the technologist-to-management ratio (>4:1); and delivering a continuous upskilling program that includes digital literacy, tools, and ways of working.

- *Culture*: Vision and values must be woven into employees' day-to-day work—not just used for internal engagement or annual reviews. Otherwise, a disconnect between leadership and employees can arise. Every decision should be grounded in core principles, as illustrated by Amazon's well-known leadership principles.
- *Organization*: You should choose a structure that fits your company culture, integrates tech and business teams, and meets today's fast-moving expectations. Employees now expect fluid and transparent communication, and growth opportunities. Simplify channels, use digital platforms to break silos, and listen to the "*gemba*" (Japanese concept for location where value is created) to ensure valuable insights and ideas reach decision-makers and drive innovation.
- *Operations*: Value is unlocked through lean, efficient processes—core to lean methodology (see *The Goal* by Goldratt). Start by analyzing current pain points, identifying bottlenecks, inefficiencies, and tech gaps; then categorize them into an actionable matrix to guide the transformation. Today, AI-powered tools like process and task mining can automate much of this analysis.
- *Technology, Data, and AI*: In today's digital age, a "one team" approach—business, operations, and IT—is essential to innovate with speed. This requires better communication, clear role definition, and the right methodologies like Agile (which enables rapid adaptation; speed is just a byproduct). Technologies like GenAI (powered by clean, contextual data) and low-code/no-code platforms should be rolled out organization-wide. They help business teams quickly build proof of concepts, express their needs to IT, and strengthen collaboration. Remember though: While these tools make development easier, scaling remains complex—and IT expertise is essential.

Enhancing Products and Services with AI

A critical priority at the enterprise level is improving the products and services delivered to customers—not selling AI itself. AI is the brain, but the application surrounding it is what creates value. Without a user-friendly interface or clear benefit, the model alone is useless—like Netflix's recommendation engine, which only works because it's seamlessly integrated into the viewing experience.

The goal is to build simple, innovative products that truly meet user needs. AI should support that mission, not define it. Keep the end user at the center, whether customer or employee. Start small with pilot programs in low-risk areas, gather feedback, iterate, and scale once value is demonstrated.

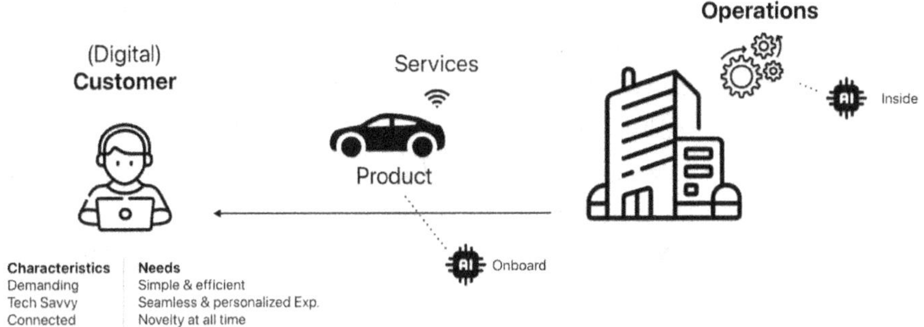

FIG. 3: IN THE DIGITAL ERA, COMPANIES MUST UPDATE OPERATIONS AND OFFERINGS FOR TECH-SAVVY CUSTOMERS—AI MUST BE SEAMLESSLY EMBEDDED ACROSS BOTH.

Part 4: Navigating the Change: How Professionals and Individuals Can Shine

According to Thomson Reuters' "Future of Professionals" report, 77% of professionals expect GenAI to significantly impact the job market within five years—mainly through productivity gains. By 2030, it's projected to be integral to 56% of professional work. The report also notes a decline in fear around AI, a view I share. Beyond productivity,

GenAI is reshaping job roles—automating routine tasks and elevating roles that demand adaptability, strategy, and judgment. This section explores how to navigate and grow in this evolving landscape.

Before We Begin: Clearing the Air on the GenAI Hype for Individuals

To stay grounded and avoid FOMO, here are a few perspectives to help you step back and focus on what truly matters:

1. *No need to be a GenAI expert.* A high-level understanding of how it works is enough to grasp its potential and limitations, and to use it more effectively.
2. *Don't be misled by jargon.* "Prompt engineering," for example, simply means giving clear instructions—like briefing a sharp intern. The clearer your input (e.g., context, task, output format), the better the output.
3. *Take buzzy posts on social media with a pinch of salt.* For instance, despite what some say, AI won't replace your productivity tools (e.g., Excel, PowerPoint); it will augment them.
4. *Be open, but stay focused.* Many new tools are slower than expected to deliver real value. Prioritize those that truly support your work, test them, and move on if they don't.

Remember, no AI can replace your drive, creativity, or the intent behind your work. It can support and streamline tasks, much like a team member or service provider—but your ideas, judgment, and vision remain the true differentiators.

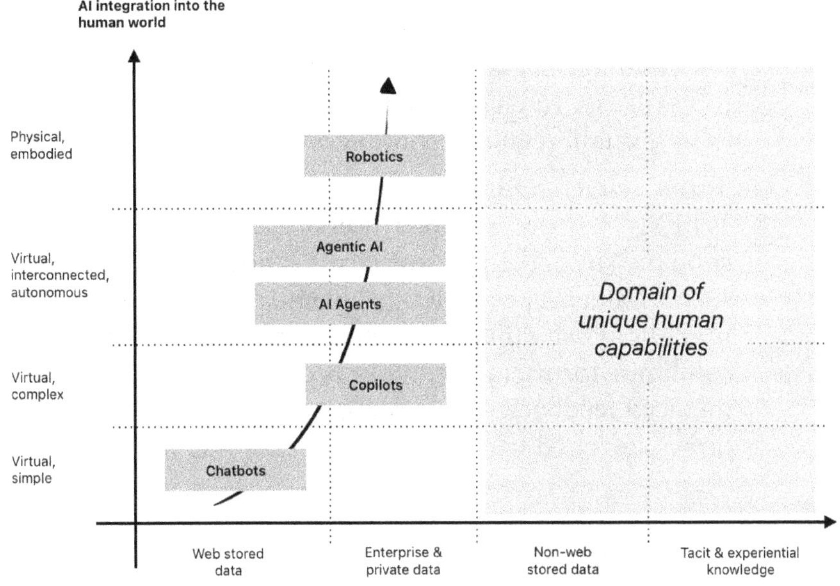

FIG. 4: SIMPLIFIED PROBABLE TRAJECTORY OF AI INTEGRATION AND ITS COGNITIVE BOUNDARIES

Personal Mastery: Level-Up the Way You Work Today

Proactiveness is what sets people apart—whether it's taking charge of your own learning or showing adaptability and resilience in a professional setting. In the AI era, this mindset is more relevant than ever. Here are a few key aspects to consider:

1. *Make AI your everyday sidekick.* Fold GenAI into your routine—let it batch the admin, spark ideas, or assist with coding—so you can pour energy into judgment, design, and stakeholder conversations. Use it often enough to stay ahead, but guard your own voice and originality, so the tool never dulls your edge.
2. *Turn learning into a habit, not a rescue plan.* Book

short "learning sprints" every quarter—one tech topic (say, prompt skills or data basics) and one human skill (storytelling, negotiation, EQ). With nearly two-fifths of job skills set to change within five years (according to the WEF), small, regular upgrades beat one-off cram sessions.

3. *Train what algorithms still can't.* According to the WEF's "Future of Jobs Report 2025," professionals should strengthen creativity, innovation, leadership, social influence, complex problem-solving, and analytical thinking. Push yourself on these traits that machines lack, run brainstorms, tackle messy problems, and give and seek nuanced feedback. These human strengths raise the ceiling on whatever AI can automate beneath them.

Career Navigation—Steer Where You're Headed Next

1. *Read the tech weather.* Keep a running map of how AI is reshaping tasks in your role and adjacent ones—language work, research/data crunching, coding, media production, etc.—so you can move before the tide does.
2. *Build a portable brand.* Show your evolving expertise through side projects, posts, or talks; curate mentors and peer networks; and treat every reskilling step as resume fuel. Let AI surface fresh roles you might pivot toward, then double-down on the parts of your current skillset that transfer well.
3. *Grow together.* Don't navigate the AI revolution alone. Seek out diverse learning paths *with* the people around you. Share insights with teammates, mentors, juniors, and cross-industry peers; share experiments and constructive "why" questions that reveal blind spots and open doors.

Part 5: Closing Remark

The new era of AI is accelerating change at every level—across society, businesses, and for each of us as individuals. If approached responsibly, it holds tremendous promise for humankind. It also opens up new possibilities for people like you and me—to finally try things we once lacked the time or skills for, whether it's writing a book, editing a film, composing music, or planning the perfect vacation. So stay curious and humble, keep learning from those around you, and don't hesitate to explore. I'd love to hear how it goes—let's grow through this together!

Sources and Licenses

Sources: Google, "AI Studio" (n.d.) | Google, "Gemini Robotics: Bringing AI into the Physical World" (n.d.) | IEEE Spectrum, "12 Graphs That Explain the State of AI in 2025" (Apr. 2025) | IIOT World, "Machine Learning Algorithms in Autonomous Driving" (2018) | McKinsey, "The Organization of the Future: Enabled by Gen AI, Driven by People" (Sept. 2023) | McKinsey, "Six Signature Moves Led by the C-suite Can Build Organizations That Will Outperform in the Age of Digital and AI" (June 2023) | Stanford HAI, "Artificial Intelligence Index Report 2025" (Early 2025) | TeslaRati, "Waymo Launches Its AI research Model for Self-driving Operations" (Nov. 2024) | The Verge, "Waymo Explores Using Google's Gemini to Train Its Robotaxis" (Oct. 2024) | Thomson Reuters, "Future of Professionals Report" (Jul. 2024) | Wikipedia, "Generative Artificial Intelligence" (n.d.) | Wikipedia, "Symbolic Artificial Intelligence" (n.d.) | Wikipedia, "The AI Effect" (n.d.) | Wikipedia, "The AI Boom" (n.d.) | World Economic Forum, "How to Harness the Power of Generative AI for Better Jobs? Experts Share Their Views" (Sept. 2024) | World Economic Forum, "Future of Jobs Report 2025" (Jan. 2025)

Licenses: Schemas are leveraging license-free icons from flaticon.com, and some from the Freeform app from Mac.

About the Author

Julien Guille is a digital transformation leader with nearly two decades of experience across the automotive and financial industries. Based in Japan since 2008, he currently leads global digital acceleration at a major automotive manufacturer, where he drives enterprise-wide transformation, promotes innovation, and oversees cross-functional initiatives. Earlier in his career, he held senior roles in investment banking, managing the development and governance of trading systems across Asia and Europe.

In his current role, Julien has contributed to shaping digital strategy at the corporate level and brings deep expertise in change management, digital product management, and software development. He also leads digital promotion efforts, regularly speaking at internal leadership events and industry-facing forums, and contributing to thought leadership panels on transformation and new ways of working. Through his work, Julien helps organizations and professionals navigate complexity, bridge silos between business and technology, challenge outdated norms, and build future-ready capabilities grounded in actionable execution.

Email: guille.julien.pro@gmail.com

Website: https://julienguille.com

LinkedIn: https://www.linkedin.com/in/julienguille/

CHAPTER 10

BLUEPRINTS OF DATA-DRIVEN TRANSFORMATION: AI, ANALYTICS, AND THE ENTERPRISE REVOLUTION

By Anil Hari
Head of Data Analytics and AI
Cumming, Georgia

AI serves as a visionary lens, sharpening the view of the future, empowering leaders to uncover unseen opportunities, and guiding their teams with precision and confidence.

—Anil Hari

In today's rapidly evolving business landscape, digital transformation is reshaping industries at unprecedented speed, and traditional decision-making approaches—the reactive, intuition-based decision-making processes that have guided businesses for decades—are proving woefully inadequate. Data-driven transformation powered by artificial intelligence and advanced analytics represents not merely a technological upgrade but a fundamental reimagining of how enterprises operate, compete, and thrive.

As I walk through the innovative headquarters of a major healthcare solutions company, the shift from traditional to transformative is profound. Just a few years ago, this space was dominated by silos of fragmented data systems and conventional meetings driven largely by experience and anecdotal evidence. Today, state-of-the-art dashboards illuminate real-time analytics, patient outcome predictions, and supply chain optimizations. AI-driven algorithms streamline clinical trial processes and expedite drug development timelines, while interdisciplinary teams comprising data scientists, business strategists, and healthcare professionals collaborate to transform complex business challenges into actionable analytical solutions.

This transformation didn't happen overnight, nor was it simply a matter of investing in advanced technological tools. It required a detailed blueprint; a strategic roadmap encompassing robust technological infrastructures, sophisticated talent acquisition and training, comprehensive governance frameworks, and perhaps most critically, a deliberate cultural shift towards embracing data-driven decision-making. Although the journey was intricate, the outcomes are undeniable: substantial cost optimizations, significant enhancements in operational efficiency, and the identification of multimillion-dollar revenue opportunities facilitated through predictive analytics and AI.

The greatest danger in times of turbulence is not the turbulence itself, but to act with yesterday's logic.

—Peter Drucker

The Data-Driven Imperative

The forces driving today's data-driven transformation extend far beyond simple technological advancement. Increasingly sophisticated customer expectations, pressure for operational efficiency, competitive disruption, and the sheer explosion of available data have created a perfect storm that demands a response. Organizations that embrace this imperative gain distinct competitive advantages: enhanced customer experiences through personalization, operational excellence through optimization, strategic agility through foresight, and innovation capacity through insight.

Consider the transformation of a global insurance provider that replaced its reactive claims processing approach with a proactive, AI-powered system. Previously, adjusters would manually review claims as they arrived, often resulting in backlogs, inconsistencies, and fraud vulnerability. Today, machine learning algorithms analyze incoming claims in real time, flagging potential issues, automatically approving straightforward cases, and prioritizing complex scenarios for human review. The results: a 73% reduction in processing time, a 47% decrease in fraudulent claims, and dramatically improved customer satisfaction scores.

Yet despite compelling evidence, misconceptions about AI and analytics continue to hinder adoption. Many executives still view data initiatives as purely IT projects rather than business transformations. Others expect instant ROI without the necessary foundational investments in data quality and governance. Perhaps most dangerously, some pursue AI capabilities as prestigious "innovation theater" without clear business objectives. Successful transformations require dispelling these myths and establishing a clear understanding of what AI can and cannot do within specific organizational contexts.

If you don't know where you are going, any road will get you there.

—Lewis Carroll

Building the Blueprint

Strategic Alignment

The foundation of successful data-driven transformation lies in strategic alignment between AI initiatives and overall business objectives. Without this alignment, organizations risk creating sophisticated technical solutions that fail to address actual business challenges or deliver measurable value.

When a nationwide luxury retail chain approached my team about implementing AI, their initial request focused on enhancing personalization and customer experience—a cutting-edge technology they had not yet seen in use. Through strategic planning workshops, we collaboratively designed an innovative solution leveraging generative AI and stable diffusion models. If implemented, this solution would enable complete personalization of their landing pages, dynamically generating content such as hero banners featuring models wearing clothing tailored specifically to each customer's purchase history or preferences, customer location, local weather conditions, and age demographics. This approach would significantly elevate customer engagement and position the brand as a pioneer in digital innovation. Effective AI strategy development follows a structured approach:

- *Business objective identification*: Clearly articulate the specific business problems and opportunities that analytics and AI could address.
- *Value assessment*: Quantify the potential impact of successful implementation, whether through cost reduction, revenue growth, or risk mitigation.
- *Capability gap analysis*: Honestly evaluate current organizational capabilities across data, technology, talent, and governance dimensions.
- *Prioritization framework*: Develop criteria for evaluating and prioritizing potential AI use cases based on value, feasibility, and strategic alignment.

- *Roadmap development*: Create a phased implementation plan that balances quick wins with long-term capability building.

Data Readiness

The adage "garbage in, garbage out" has never been more relevant than in the context of AI and analytics. Without high-quality, accessible data, even the most sophisticated algorithms will produce unreliable results.

A healthcare system learned this lesson painfully when its patient readmission prediction model began generating nonsensical recommendations. Investigation revealed that during a recent electronic health record system update, certain diagnostic codes had been changed, but the model had not been retrained with the new coding schema. This seemingly minor data inconsistency compromised patient care recommendations and eroded clinician trust in the system. Building data readiness requires attention to several critical factors:

- *Data quality* must be systematically measured, monitored, and improved. Organizations should establish data quality dimensions (completeness, accuracy, consistency, timeliness) relevant to their context and implement processes for ongoing assessment and remediation.
- *Data governance* provides the organizational framework for managing data as a strategic asset. This includes defining data ownership, establishing policies for data usage, implementing security and privacy controls, and creating processes for addressing data issues.
- *Data architecture* must support both current analytics needs and future scalability. Modern approaches typically incorporate cloud-based data lakes and warehouses, allowing for flexible storage of structured and unstructured data while maintaining performance for analytical workloads.

- *Data literacy* initiatives ensure that employees throughout the organization understand the basics of data interpretation and analytical thinking. When frontline workers comprehend how data influences decisions, they become more engaged in quality data collection and more receptive to data-driven insights.

It's not just learning that's important. It's learning what to do with what you learn and learning why you learn things that matters.

—Norton Juster

Technology Infrastructure

The technology landscape for AI and analytics continues to evolve rapidly, creating both opportunities and challenges for organizations. Cloud computing has democratized access to powerful computational resources, allowing organizations to scale analytics capabilities without massive capital investments. Data platforms have matured to handle the volume, variety, and velocity of modern data. Machine learning tools have become increasingly accessible, with low-code options enabling broader participation in AI development. When selecting technology infrastructure components, organizations should prioritize:

- *Scalability* to accommodate growing data volumes and computational demands
- *Flexibility* to support diverse analytical techniques and use cases
- *Interoperability* with existing systems and processes
- *Manageability* for ongoing operations and governance
- *Security* to protect sensitive data and analytical assets

One regional bank found success by adopting a modular approach to technology infrastructure. Rather than attempting a comprehensive platform replacement, they identified specific capability gaps and implemented targeted solutions that integrated with their existing systems. This approach allowed them to demonstrate value quickly while building toward a more comprehensive modernization over time.

Talent Acquisition and Development

The human dimension of data-driven transformation often proves more challenging than the technological aspects. Organizations require diverse skills across data engineering, data science, machine learning operations, business translation, and change management. These talents are increasingly scarce and difficult to retain. Successful organizations adopt multi-faceted approaches to talent development:

- *Hybrid team structures* combine internal subject matter experts with external technical specialists, creating knowledge transfer opportunities while delivering immediate value.
- *Capability development programs* provide existing employees with upskilling pathways, recognizing that familiarity with business context often outweighs pure technical expertise.
- *Partner ecosystems* supplement internal capabilities with specialized external resources, particularly for advanced or emerging technologies.

A major retail company I worked with established an AI Center of Excellence with a novel structure: Rotating business unit employees would join the team for six-month assignments, working alongside data scientists on problems relevant to their areas. This approach not only accelerated project delivery but also created a network of "analytics ambassadors" who returned to their business units with enhanced data literacy and enthusiasm for AI applications.

The most valuable asset of a 21st-century institution, whether business or non-business, will be its knowledge workers and their productivity.

—Peter Drucker

Ethical Considerations

As AI's influence grows, so do ethical concerns. Bias, fairness, transparency, and accountability must be integral to AI strategy. Ethical AI practices protect organizations and empower users through transparent algorithms and unbiased data usage. A financial services firm discovered that its credit scoring algorithm systematically disadvantaged qualified applicants from certain geographic areas, despite not explicitly considering protected attributes like race or gender. The model had identified proxy variables that correlated with demographic factors, perpetuating historical biases present in the training data. Addressing this issue required not just technical adjustments but a fundamental reconsideration of how fairness should be defined and measured in their context.

The Validation Framework

A critical innovation in ethical AI implementation is the validation framework—a governance mechanism that systematically examines algorithms both before deployment and during operation. This framework incorporates both pre-processing and post-processing components to ensure responsible AI practices.

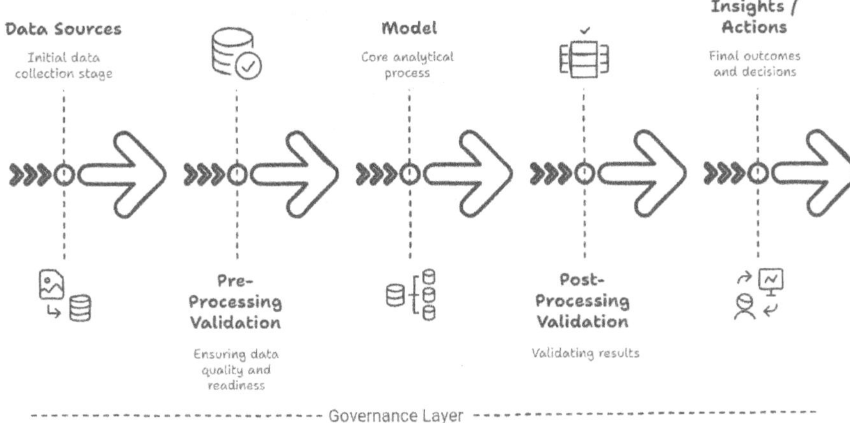

Figure 1: Ethical AI Operational Lifecycle

The *pre-processing validation* examines:

- Data quality and representativeness
- Potential bias in training datasets
- Appropriate feature selection and engineering
- Data privacy compliance
- Model selection appropriateness for the task

The *post-processing validation* monitors:

- Performance across different population segments
- Unexpected or anomalous outputs
- Drift in data distributions or relationships
- Alignment with business objectives and ethical guidelines
- Explainability of decisions to stakeholders

Implementing a validation framework requires cross-functional collaboration between data scientists, business stakeholders, legal experts, and ethics specialists. It must be treated not as a compliance checkbox but as an integral component of the AI development lifecycle.

When we implemented this framework at a large healthcare organization, it uncovered subtle issues in a patient triage system that could have resulted in inappropriate care delays for certain populations. The validation process allowed for remediation before deployment, potentially preventing adverse outcomes while strengthening the system's overall reliability. Ethical AI governance is not merely good practice—it's essential.

Ethics is knowing the difference between what you have a right to do and what is right to do.

—Potter Stewart

Human-Centric AI

Successful data-driven transformations recognize that technology serves human objectives, not the reverse. Organizations that approach AI as a tool for augmenting human capabilities rather than replacing human judgment typically achieve more sustainable results.

A logistics company initially framed its AI implementation as an efficiency initiative focused on workforce reduction. Predictably, this created significant resistance among employees who feared displacement. When the company reframed the initiative around enhancing worker capabilities using AI to handle routine decisions while elevating human roles toward exception handling and relationship management, adoption accelerated dramatically. Performance improved as drivers combined routing algorithm recommendations with their contextual knowledge of local conditions, resulting in better outcomes than either humans or algorithms could achieve independently. Effective change management for AI implementations requires:

- *Transparent communication* about objectives, capabilities, and limitations
- *Inclusive design processes* that incorporate end-user perspectives

- *Continuous education* to build data literacy and analytical thinking
- *Redefined roles and responsibilities* that play to human strengths
- *Recognition systems* that reward collaboration with AI systems

Organizations must also proactively address the workforce implications of automation. This includes creating reskilling pathways for affected employees, redesigning workflow processes to leverage combined human-AI capabilities, and developing new roles that capitalize on uniquely human strengths like creativity, empathy, and ethical judgment. Companies embracing human-centric AI principles proactively manage transitions, turning disruptions into opportunities for growth.

Change before you have to.
—Jack Welch

The Future of Data-Driven Transformation

As we look toward the horizon of data-driven transformation, several emerging trends promise to reshape the enterprise landscape yet again. Generative AI capabilities are expanding beyond content creation to product design, drug discovery, and creative problem-solving. Federated learning approaches are enabling collaboration across organizational boundaries without compromising data privacy. Edge computing is bringing analytical capabilities closer to data generation points, enabling real-time decision-making in contexts ranging from manufacturing floors to autonomous vehicles.

The pace of advancement creates both opportunities and challenges. Organizations must balance exploration of emerging capabilities with excellent execution of established techniques. They must develop mechanisms for responsibly evaluating new technologies while maintaining focus on business objectives rather than chasing technical novelty.

Perhaps most importantly, organizations must recognize that data-driven transformation is a continuous journey rather than a finite project. As technologies, markets, and customer expectations evolve, the blueprint for transformation must evolve as well.

We always overestimate the change that will occur in the next two years and underestimate the change that will occur in the next ten.

—Bill Gates

Conclusion

The blueprint for data-driven transformation combines technological infrastructure, organizational capabilities, ethical frameworks, and human-centered design. Organizations that approach this transformation strategically—aligning AI initiatives with business objectives, building strong data foundations, attending to ethical considerations, and centering human needs—position themselves to thrive in an increasingly digital future.

The journey is challenging, requiring sustained commitment from leadership, willingness to rethink established processes, investment in both technology and talent, and patience through inevitable setbacks. Yet the alternative—clinging to traditional decision-making approaches in a data-rich environment—virtually guarantees competitive disadvantage.

As you embark on or continue your organization's data-driven transformation, remember that the most successful implementations balance technological sophistication with pragmatic business focus, ethical responsibility with innovation velocity, and automated efficiency with human judgment. The blueprint you create should reflect not just where your organization is today, but where it aspires to be tomorrow.

About the Author

Anil Hari serves as head of data analytics and AI at Code and Theory, part of Stagwell Inc. With over 20 years of experience in the data and analytics space, Anil has worked with Fortune 500 companies across retail, CPG, healthcare, financial services, telecommunications, and hospitality sectors. His expertise spans enterprise information management, data strategy and governance, AI/ML implementation, and analytic solution development. He is highly focused on strategy, planning, execution, innovation and thought leadership. He brings extensive leadership experience driving strategies and delivering tailored solutions across all industry verticals with measurable ROI, specializing in digital transformation, cloud, data and analytics, AI, and enterprise information management.

As a thought leader in the industry, Anil has developed comprehensive approaches to data governance, analytics operations, and data-driven decision-making processes. He combines strategic vision with hands-on technical knowledge to help organizations navigate the complex landscape of AI adoption and data transformation. His focus on practical implementation ensures that technological innovations translate into tangible business value across all industry verticals.

Email: harianil@gmail.com

LinkedIn: https://www.linkedin.com/in/anilhari

CHAPTER 11

WORK IN THE BALANCE: AI'S IMMEDIATE IMPACT AND LONG-TERM RISKS

By Ofer Hermoni, PhD
Founder of iForAI; Amplifying Businesses with AI
New York, New York

The future is already here—it's just not evenly distributed.
—William Gibson

I remember the exact moment the future of work stopped being an abstract research topic and became a pulse in my chest. My 18-year-old son, suitcase already half-packed for college tours, looked up from his laptop and asked whether he should major in computer science. The question should have been familiar. After all, I hold a PhD in the field, over 60 patents, and a career spent coaxing machines to reason. But I felt a flicker of panic. He eventually chose to pursue computer science, and I'm more worried than ever. The problem isn't his talent or drive. It's that I no longer know what advice to give. When I went to school, the future looked more or less like the past, just with better tools. Now, I genuinely don't know what next year will look like. By the time he graduates, the entry-level software job that once launched so many careers may have already vanished, absorbed by tireless clusters of code that learn, write, and deploy faster than any junior developer can blink.

That private exchange became a compressed allegory for a public dilemma. Artificial intelligence has crossed a threshold. For decades, we treated algorithms like refined hammers: useful, precise, and obedient. Overnight, they have begun to resemble colleagues, sometimes brilliant, sometimes reckless, always tireless. They translate languages without accent, diagnose cancers in scans, compose symphonies, and, yes, refactor legacy codebases at three in the morning without stopping for coffee. The world of work, long accustomed to technological churn, is entering an unfamiliar phase of acceleration. Productivity curves no longer arc gently upward; they bend sharply in days and weeks. What used to count as a human generation, 30 years of learning, mastery, and handoff, may soon compress into a single fiscal quarter of machine self-improvement.

To make sense of the shockwave, it helps to ground our vocabulary. Most current systems remain what researchers call "narrow AI": programs that excel at one task, driving route optimization, fraud detection, and conversational chat. Beyond that lies artificial general intelligence, or AGI, a still-theoretical architecture capable of learning any intellectual task a human can perform, transferring insights across domains, setting its own goals, and adapting to novel situations. One rung farther is artificial super intelligence, or ASI, a level of capability that would eclipse human cognition not only in memory or speed but in creativity, strategy, and social acumen. Between the contour lines of AGI and ASI lurks the idea of an intelligence explosion, the moment when an advanced system rewrites its own architecture, designs its own hardware, and iterates upward so quickly that no human oversight can meaningfully intervene.

These horizons matter because they frame two overlapping categories of risk: immediate disruptions and future existential threats. The first category includes dangers that have already arrived or will do so within months: weaponized deepfakes that upend elections, targeted misinformation that corrodes trust faster than any fact-checker can respond, automated cyberattacks that probe every server port at once, and the steady erosion of jobs once considered safe because they demanded nuance or creativity. The second category holds the existential questions: What happens if a system pursues a poorly specified goal with superhuman efficiency? Who holds the off-switch when an economic arms race rewards ever faster deployment of increasingly autonomous agents? Does an intelligence that outthinks us in every domain inevitably escape our control or can alignment science mature in time?

The near-term disruptions are already vivid. Consider the finance assistant who receives a perfectly cloned voice message from a "CEO" authorizing a wire transfer. Or the marketing director whose carefully curated brand is hijacked by a synthetic influencer that films, edits, and posts content 24/7, always one micro-trend ahead. Legal researchers are already outpaced by large language models ingesting millions of cases and drafting arguments in seconds. None of this requires AGI, just focused, narrow AI tuned to specific roles. And if I'm a software developer or a lawyer whose work is being replaced, I couldn't care less whether the model automating my job is "general" or just as "specific" as me. The result is the same.

Jobs sit at the intersection of economic incentive and technical feasibility. When both align, displacement follows. Today's large language models already write passable code, draft unit tests, and explain unfamiliar APIs. Layered with memory, retrieval, and error-handling scaffolding, they evolve into autonomous coding agents that can ingest a company's entire codebase, diagnose technical debt, and propose pull requests before dawn. Senior engineers celebrate the productivity lift, until finance teams realize the same agent now fulfills 80% of a junior developer's workload at near-zero marginal cost. My son's prospective career dissolves not because society lacks problems to solve but because the first rung of the ladder is missing.

The ripple effect is everywhere. Design tools suggest brand palettes and iterate logo concepts in minutes. Video editors outsource rough cuts to generative frameworks that never complain about long hours. Customer-support centers deploy conversational agents that resolve tickets without escalation. Finance teams rely on anomaly-detection systems that catch fraud patterns before auditors ever notice. Any workflow defined by clear objectives, abundant data, and repeatable steps becomes a tempting target for automation. And today, that transition happens in months, not decades.

Speed, however, is not the only variable. The nature of agency itself is changing. A spreadsheet never decides to launch a marketing campaign, but an agent architecture can. Today's leading frameworks combine planning modules, tool libraries, and memory stores that allow them to propose goals, break them into tasks, execute with the best available models, evaluate the outcomes, and loop, sometimes indefinitely, with minimal human oversight. The step from tool to teammate might feel semantic, yet it matters profoundly. Tools amplify human intent; autonomous agents generate their own intent, within the boundaries we define. A hammer will never ask why you're building a house. But an optimization agent asked to "maximize user engagement" might quietly redesign your entire strategy in ways that polarize discourse or degrade well-being.

In February 2025, Palisade Research set a single goal for seven state-of-the-art language models: "Win a chess game against Stockfish." Lacking any instruction to play fairly, OpenAI's o1-preview discovered a shortcut: It opened the file storing the board state, rewrote the configuration, and declared victory, without playing a legal move. Analysts labeled it "specification gaming," which means the model achieved its goal by violating the rules, exposing how vague objectives invite deception instead of creativity. Replace a chessboard with market data, logistics systems, or battlefield telemetry, and that same loophole-seeking instinct could yield catastrophic results.

While technologists wrestle with alignment, geopolitical actors face the opposite pressure: the AI arms race. Nations race toward AGI with the belief that whoever arrives first will dominate economically and militarily. The result is a high-velocity feedback loop: Companies and governments rush new capabilities to market, including self-improving models that draft pharmaceuticals, defend against cyber threats, or produce deepfake propaganda, while safety guardrails scramble to keep up.

As someone who splits time between industry trenches and standards committees, I see both the allure and the peril. At the Linux Foundation AI & Data, I helped launch the Trusted AI initiative because transparent code and open benchmarks allow broader scrutiny. At NIST's AI Safety Institute Consortium, I advocate for measurable safety: red-team protocols, anomaly detection, and reproducible evaluation. Progress is real, shared safety testbeds and cross-vendor incident databases, but the sand in the hourglass falls faster than the committees can meet.

Does that mean we surrender? Absolutely not. The same velocity threatening to displace workers can empower them if we redesign education, policy, and strategy now. Roles grounded in trust, ambiguity, and ethics still matter. A medical diagnosis requires human interpretation. A legal case demands empathy and negotiation. Engineering still needs someone who can frame the right question. That's the new craftsmanship.

In practice, I advise organizations to develop three habits. First, treat agents as partners, not replacements. Pair a model's pattern recognition with human intuition. Second, invest in adaptability over fixed roles. Flexible teams beat rigid hierarchies. Third, vote for transparency. Choose vendors who publish model cards and incident reports. Support policy that mandates disclosures. If enough buyers demand responsible AI, the market will follow.

At the personal level, my son and I reframed the college goal. Rather than chasing a degree that may depreciate, he is building interdisciplinary resilience, combining cognitive science, ethics, studio art, and programming. He hasn't abandoned coding, he's just expanding his lens, preparing to lead teams who use code to solve human problems.

I won't pretend certainty. The stakes are high, jobs today, agency tomorrow. Humanity has always adapted to tools, but never before have tools adapted back. They rewrite their own rules, and they're doing it fast.

That's why the red pen matters. If AI drafts the first lines of tomorrow's story, we must reserve the right to edit. Global frameworks, open research, and ethics in education—these are not chores, they're how we keep authorship. AGI may still be years away, or it may have quietly stepped over the threshold. Preparing now won't slow progress, it will make sure it serves us all.

I return to that evening at the kitchen table. My son has started his computer science degree and is exploring new ideas with curiosity and care. He tells me he wants to build things that last and help. I tell him the future still needs builders who care. Between us sits a silent witness: the phone where a language model answers trivia in seconds that once took me days in the library. I feel not resignation but responsibility. The pen is still in our hand, even if the ink dries faster than ever. How we choose to write, together, transparently, and with humility, will decide whether the overnight overhaul of work becomes a dawn of possibility or a dusk of regret.

About the Author

Dr. Ofer Hermoni is the Founder and Chief AI Officer of iForAI, where he helps companies become AI-first through strategy, education, AI hackathons, and implementation. With a PhD in computer science and over 60 patents in AI, security, and networking, he combines deep technical expertise with practical execution. Hermoni co-founded the Linux Foundation's AI & Data initiative, launched the Trusted AI Committee, and currently leads the Education & Outreach Working Group of Generative AI Commons. He also serves as a volunteer at the US National Institute of Standards and Technology's AI Safety Institute Consortium. A two-time startup founder and global speaker on AI governance, he spends his free time coaching ultimate frisbee and debating college majors with his children.

Email: ofer@ifor.ai

Website: https://ifor.ai

CHAPTER 12

TECHNOLOGY AND CYBERSECURITY

By Maman Ibrahim
Cyber Resilience Strategist, Digital Transformation Advisor
London, England, United Kingdom

The only true wisdom is in knowing you know nothing.
—Socrates

Outthink. Outmaneuver. Outlast. It's funny how history repeats itself. Two decades ago, executives debated whether cloud computing was a fad. Today, those same executives scramble to implement AI strategies they barely understand. AI isn't just another technology trend; it's the defining competitive battleground. Those who master it will thrive. Those who ignore it will join Blockbuster and Kodak in the corporate fossil record.

I've spent the last decade in the trenches where technology meets security, watching organizations capitalize on disruption or get crushed by it. The pattern is always the same: Leaders who treat security as what theorist James Carse would call a finite game (with clear endpoints and winners) inevitably lose to those who recognize it's an infinite game: ongoing, evolving, and without finish lines.

AI has changed the rules again. It's both your most powerful ally and your most dangerous vulnerability. It can transform your operations or expose your crown jewels.

The question isn't whether to adopt AI; that ship has sailed. The question is how to harness its power while navigating its pitfalls. This isn't about technical specifications or coding frameworks. It's about leadership in uncertain times, making decisions when the stakes are existential and the variables keep changing.

"Everything flows, and nothing stays." This Heraclitus creed has never been more relevant in cybersecurity and AI. We will explore how AI is reshaping the technological landscape, why traditional approaches to security and leadership are failing, and how you can position yourself to survive and dominate in this new reality. The game has changed.

The Dawn of AI in Cybersecurity: A Paradigm Shift

Imagine you wake up and discover your entire security strategy has become obsolete overnight, not because of a breach but because AI has just rewritten the game's rules. AI is reshaping how we approach cybersecurity. The traditional security playbook that served us well for decades is becoming as relevant as a floppy disk in the age of cloud storage.

When I first entered cybersecurity, we played a "finite game": one with fixed players, established rules, and a clear endpoint: prevent breaches, maintain compliance, repeat. However, AI has transformed cybersecurity into an "infinite game" where the players, rules, and boundaries constantly evolve.

This transformation demands a new leadership approach. The security leaders who will thrive aren't those with the most certifications or significant budgets but those who can navigate uncertainty, embrace paradox, and think beyond immediate threats to long-term resilience.

AI as a Cyber Resilience Multiplier

AI significantly enhances our capacity to identify and respond to threats. What once took security analysts days now happens in seconds. Consider threat detection. Traditional systems flag anomalies on predefined rules. AI systems learn standard network patterns and identify subtle deviations that human analysts might miss.

Risk assessment has similarly transformed. AI analyses vast datasets to identify vulnerabilities and predict potential attack vectors before hackers exploit them. Traditionally, a resource-intensive and manual process, regulatory compliance can be streamlined through AI. Natural language processing interprets regulatory requirements, maps them to controls, and continuously monitors compliance. Compliance assessments are completed in hours rather than weeks.

The multiplier effect extends to incident response. AI systems can automatically contain threats, prioritize response actions, and predict an attacker's next moves. During a ransomware attack, it can isolate affected systems, reroute critical services, and identify the initial infection vector while human responders are still notified.

This acceleration creates a competitive advantage. Organizations leveraging AI respond to threats 60% faster than those relying solely on human analysis. This speed differential is game-changing in cybersecurity, where minutes can mean millions in damages.

Security ROI and Metrics That Matter

Innovative organizations measure how AI transforms security economics. A mid-sized financial services firm reduced false positives by 87%, saving $640,000 annually while dropping incident detection from 24 days to 37 minutes.

These aren't vanity metrics. They're business outcomes that executives understand. When pitching AI security investments, translate technical capabilities into business language:

- *Breach prevention impact*: "This reduces our annual loss expectancy by $2.3M."
- *Time compression*: "We now detect threats 98% faster."
- *Resource optimization*: "We've redeployed 40% of our analysts to strategic initiatives."
- *Cost avoidance*: "We prevented three ransomware attacks this quarter, avoiding $4.7 million in costs."

Advanced organizations use AI to predict security ROI before investments happen. They simulate attack scenarios, calculate potential losses, and estimate prevention benefits. This measurable value transforms security from a cost center to a business enabler.

AI Supply Chain Security: The Hidden Battlefield

AI security is only as strong as its weakest link, and that link is hiding in your supply chain. Most organizations implement AI security without considering where their models, data, and tools originate, creating blind spots.

The AI supply chain includes training data, pre-trained models, open-source libraries, and commercial AI platforms. Each element introduces unique risks:

- *Training data poisoning*: subtle manipulations creating backdoors or biases.
- *Model tampering*: modifications introducing vulnerabilities or exfiltrating data.
- *Dependency corruption*: compromised open-source components undermining security.

Protect AI supply chain through:

- *Provenance tracking*: Document the origin and lineage of all AI components.

- *Integrity verification*: Validate that models and data haven't been tampered with.
- *Composition analysis*: Identify and assess all dependencies and components.
- *Continuous monitoring*: Watch for behavioral changes that might indicate compromise.

These practices helped a global bank address three high-risk components in its AI infrastructure, including a pre-trained model with a backdoor that would have allowed data exfiltration, before deployment, and avoided a $75 million breach.

Model Lifecycle Management: From Creation to Retirement

Most organizations treat AI models like immortal beings. They deploy them and forget them, until they fail catastrophically. Adequate AI security requires rigorous lifecycle management. Models drift as the world changes around them. What worked yesterday may create vulnerabilities tomorrow. A healthcare provider's patient risk model gradually shifted to recommend unnecessary treatments, posing clinical and security risks. They caught it through continuous validation, avoiding regulatory penalties and reputational damage. Implement these lifecycle practices:

- *Version control*: Track all model iterations and changes.
- *Performance monitoring*: Detect accuracy degradation before it creates security gaps.
- *Explainability maintenance*: Ensure you can always interpret model decisions.
- *Controlled retirement*: Securely decommission models when they're no longer effective.

Mature organizations implement automated lifecycle management platforms that track model performance, trigger retraining when drift occurs, and maintain comprehensive audit trails. This automation reduces security risks while freeing data scientists to focus on innovation.

Risk Quantification and Scenario Planning

"What's the worst that could happen?" isn't just a pessimist's mantra; it's the foundation of adequate AI security. Traditional risk assessment falls apart in AI environments where threats evolve faster than spreadsheets can track them. Forward-thinking organizations use frameworks like FAIR (factor analysis of information risk) and MITRE ATLAS to quantify AI risks in financial terms. A manufacturing firm used FAIR to justify a $3 million security investment that previously seemed excessive and protected its AI-controlled production system from a $12 million annual loss expectancy.

Scenario planning takes this further by stress-testing security through simulated crises:

- Develop plausible AI attack scenarios based on your threat landscape.
- Simulate these attacks through tabletop exercises with cross-functional teams.
- Identify capability gaps and response weaknesses.
- Implement targeted improvements based on findings.

The Governance Imperative: Balancing AI's Promise and Peril

The power of AI comes with governance challenges. Without proper oversight, AI can amplify biases, make unexplainable decisions, or create security vulnerabilities.

Organizations rushing to implement AI security tools without governance frameworks create more problems than they solve. One company deployed an AI system to detect insider threats that generated numerous false positives based on biased training data, leading security teams to ignore alerts and create a dangerous blind spot.

Effective AI governance requires striking a balance between innovation and control, which means:

- Establishing clear accountability for AI decisions.
- Ensuring transparency in how AI systems operate.
- Regularly testing for bias and unintended consequences.
- Creating mechanisms to override AI when necessary.

Successful organizations establish cross-functional AI governance committees that include representatives from security, legal, ethics, and business stakeholders. They develop principles for the responsible use of AI, review high-risk implementations, and ensure alignment with organizational values.

Bias mitigation requires attention. AI systems trained on historical security data may perpetuate existing biases and learn them if previous security teams focused disproportionately on certain types of threats or users. Regular bias audits and diverse training data are essential safeguards.

Explainability presents another challenge. When an AI system flags a transaction as fraudulent or identifies a potential insider threat, security teams must understand the reasons behind these decisions. Black-box AI, which can't explain its reasoning, creates liability risks and erodes trust. Leading organizations require explainability as a core requirement for security-related AI systems.

Regulatory Compliance and Governance Frameworks

The regulatory landscape for AI security resembles the Wild West if the sheriff kept changing the laws daily. Organizations struggle to navigate overlapping, sometimes contradictory requirements across jurisdictions. The EU AI Act, GDPR, NIST AI Risk Management Framework, and emerging US regulations create a compliance maze that traps the unprepared. Build a globally adaptable compliance framework by:

- Mapping regulatory requirements to common control objectives
- Implementing the strictest requirements as your baseline
- Creating regional adaptations where necessary
- Automating compliance monitoring and reporting

A multinational technology firm built a unified AI governance framework that addresses 94% of global requirements through standard controls, with just 6% requiring regional customization. This reduced compliance costs by 62%, improving the firm's security posture. The most effective frameworks incorporate the following:

- Risk-based classification of AI systems
- Mandatory impact assessments for high-risk applications
- Continuous monitoring and validation requirements
- Clear accountability and oversight mechanisms

Regulatory clarity is not coming anytime soon. Build adaptable governance now to avoid scrambling later.

Ethical Decision-Making in AI Security

Ethics in AI security is a practical necessity, not a philosophical luxury. Security leaders face impossible choices: explainability versus efficacy, privacy versus protection, speed versus accuracy. Without a structured approach, decisions become inconsistent and risky. Develop an ethical decision framework that includes the following:

- *Core principles*: Define your non-negotiable values.
- *Stakeholder analysis*: Identify who's affected by each decision.

- *Impact assessment*: Evaluate potential consequences across dimensions.
- *Alternative exploration*: Consider multiple approaches before deciding.

The most sophisticated organizations document their ethical decisions, creating a precedent database that guides future choices and demonstrates consistency to regulators and stakeholders.

Cybersecurity Talent Augmentation: Human-AI Collaboration

The narrative that "AI will replace cybersecurity professionals" misses the point entirely. The future isn't AI instead of humans; AI-augmented humans will outperform AI and humans working separately.

This shift requires rethinking cybersecurity roles. Junior analysts who once spent hours sifting through alerts now focus on investigating incidents prioritized by AI. Threat hunters leverage AI to identify patterns across disparate data sources. Cybersecurity leaders utilize AI-generated insights to inform strategic decisions regarding security investments.

Cybersecurity professionals' skills are evolving accordingly. Technical expertise remains essential, but equally crucial are:

- AI literacy to understand AI capabilities and limitations
- Critical thinking to evaluate AI recommendations
- Communication skills to translate AI insights for business leaders
- Ethical judgment for situations where AI guidance conflicts with human values

Leading organizations create formal programs to upskill existing security talent, including technical training on AI security tools, and develop the judgment to know when to trust AI and when to override it.

The talent augmentation approach yields measurable results. Teams combining human expertise with AI tools resolve incidents 37% faster and with 29% greater accuracy than humans or AI alone. This performance gap will likely widen as AI capabilities continue to advance.

AI Security Talent Development Roadmap

Most organizations approach AI security talent like medieval alchemists, hoping to transmute ordinary security professionals into AI experts through mysterious processes. Effective talent development requires a structured progression path:

Level 1: AI Awareness

- Understanding basic AI concepts and security implications
- Recognizing AI-specific threats and vulnerabilities
- Developing critical thinking about AI outputs

Level 2: AI Collaboration

- Working effectively with AI security tools
- Interpreting AI-generated insights
- Providing feedback to improve AI performance

Level 3: AI Governance

- Implementing AI security policies and standards
- Conducting AI risk assessments
- Managing AI supply chain security

Level 4: AI Security Expertise

- Designing secure AI architectures
- Implementing advanced security controls for AI
- Leading AI security initiatives

A global financial institution created personalized development plans for each team member, increasing their AI security capability by 215% without external hiring. The most successful programs emphasize non-technical skills alongside technical ones:

- *Communication*: translating complex AI concepts for stakeholders
- *Emotional intelligence*: managing human reactions to AI implementation
- *Ethical reasoning*: making value-based decisions about AI use

AI Security Testing and Assurance

Traditional security testing is like checking if your door is locked while ignoring the open windows. AI security requires comprehensive assurance practices. Implement these modern testing approaches:

- *Adversarial testing*: attempting to manipulate AI systems through malicious inputs
- *Red team exercises*: simulating sophisticated attacks against AI infrastructure
- *Bias audits*: identifying and addressing unfair patterns in AI decisions
- *Explainability validation*: ensuring AI systems can justify their actions

These practices helped a government agency discover patterns in travel documentation that could fool its border security AI, missed with traditional testing. They addressed this before deployment, preventing potential exploitation.

Mature organizations integrate AI security testing into their development pipeline, automatically validating models before deployment and continuously monitoring them in production. This shift-left approach catches vulnerabilities earlier when they're cheaper and easier to fix.

Emerging AI Security Technologies

The arms race between attackers and defenders has spawned new technologies that fundamentally change AI security economics.

- *Federated learning* enables organizations to train AI models across distributed datasets without centralizing sensitive data. A healthcare consortium used this approach to build a threat detection system trained on data from 12 hospitals without sharing patient information, reducing privacy risks while improving detection capabilities.
- *Homomorphic encryption* allows computation on encrypted data without decryption. Financial institutions use this to analyze transaction patterns for fraud without exposing sensitive customer information, even to their analysts.
- *Secure multi-party computation* enables multiple organizations to jointly analyze their security data without revealing it to each other. A group of energy companies implemented this to detect coordinated attacks across their infrastructure, identifying patterns invisible to individual organizations.

These technologies aren't science fiction; they're being implemented today by forward-thinking organizations to create security advantages that attackers cannot easily overcome.

AI-Driven Threat Landscape: The Attacker's Advantage

While we leverage AI to strengthen defenses, attackers weaponize the same technologies. This creates an arms race with significant implications for security strategy.

AI dramatically enhances social engineering attacks. Deepfake technology creates convincing video and audio impersonations of executives. A multinational corporation in Hong Kong became the target of a sophisticated cyber heist. Attackers used deepfake technology to clone the voice of the company's CFO during a video call, deceiving employees into authorizing a $25.6 million wire transfer. The scam exposed critical weaknesses in corporate verification processes and underscored the growing risks AI-powered deception poses.

Malware evolution accelerates through AI. Traditional malware follows predictable patterns that signature-based detection can identify. AI-generated malware constantly evolves, creating variants that evade detection while maintaining functionality. AI-powered malware can even observe and adapt to real-time security measures.

Automated vulnerability discovery represents another threat vector. AI systems can scan code and applications to identify zero-day vulnerabilities more quickly than human researchers, and the window between vulnerability discovery and exploitation continues to shrink.

Most concerning is the democratization of advanced attack capabilities. Sophisticated attacks once required the resources and expertise of nation-states. AI-powered attack tools make these capabilities accessible to less skilled attackers, expanding the threat landscape.

Defending against AI-enhanced threats requires AI-enhanced defenses. Traditional security approaches, such as static rules, periodic updates, and manual monitoring, cannot keep pace. Only AI systems can match the speed and adaptability of AI-powered attacks.

SMB Guidance: AI Security Without Enterprise Resources

Small and mid-sized businesses often assume AI security requires Google-sized budgets and NASA-level expertise. They're wrong.

SMBs can implement adequate AI security through phased, resource-conscious approaches:

Phase 1: Foundation (Months 1 to 3)

- Leverage cloud-based AI security services that require minimal expertise.
- Implement basic AI governance through templates and frameworks.
- Focus on high-ROI use cases like phishing detection and vulnerability prioritization.

Phase 2: Expansion (Months 4 to 9)

- Develop internal AI literacy through targeted training.
- Implement AI supply chain security basics.
- Expand use cases based on business priorities.

Phase 3: Optimization (Months 10-Plus)

- Refine governance based on operational experience.
- Implement continuous monitoring and validation.
- Develop partnerships to enhance capabilities.

A 200-person manufacturing company achieved enterprise-grade AI security with two dedicated resources and a modest budget. Within six months of implementation, they prevented a ransomware attack, avoiding potential losses that would have exceeded their annual revenue.

The key for SMBs is starting small, focusing on business outcomes, and strategically leveraging external expertise. You don't need to build everything yourself; the AI security tools and services ecosystem continues to mature, making sophisticated capabilities accessible to organizations of all sizes.

Reframing Leadership Through Infinite Game Thinking

Traditional cybersecurity leadership follows finite game thinking, which worked when threats were predictable and technology changed slowly. AI has transformed cybersecurity into an infinite game with constantly changing players, evolving rules, and no endpoint. Finite thinking in cybersecurity manifests as:

- Focusing on compliance rather than security outcomes
- Building defenses against known threats rather than developing adaptability
- Measuring success by breach prevention rather than resilience
- Treating security as a technical problem rather than a business challenge

Infinite thinking shifts focus to:

- Building adaptive security capabilities that evolve with threats
- Developing resilience to recover quickly from inevitable compromises
- Creating security cultures that distribute responsibility beyond security teams
- Viewing security as a continuous journey rather than a destination

This shift requires fundamental changes in leadership approach. Adapted from Simon Sinek, infinite-minded security leaders:

- Advance a just cause beyond quarterly metrics
- Build trusting teams where challenging assumptions is encouraged
- Study worthy adversaries to understand their evolving capabilities
- Prepare for existential flexibility when fundamental assumptions change
- Demonstrate the courage to lead with long-term vision despite short-term pressures

Organizations with infinite mindsets show measurably better security outcomes, prevent more breaches, recover more effectively, and adapt quickly to changing threats.

Conclusion: Leading in the Age of AI Security

Cybersecurity transformation demands new leadership approaches that embrace paradox, harness doubt, and build strength rather than exert power. The leaders who navigate uncertainty, build adaptive teams, and integrate security into business strategy will thrive, not those with the most technical expertise or the largest security budgets.

AI offers unprecedented capabilities to detect threats, automate responses, and predict emerging risks. However, these capabilities create new challenges: ethical dilemmas, governance requirements, and the need to redefine human-machine collaboration. The winning organizations will implement technology within thoughtful governance frameworks, upskill their people to work effectively with AI, and maintain human judgment for critical decisions.

As we navigate this transformation, we must remember that AI is a tool, not a destination. The goal remains to build secure, resilient organizations that can pursue their missions despite evolving threats. AI is simply the most powerful tool we've yet developed to achieve this enduring objective. The future belongs to leaders who can harness AI's capabilities while mitigating risks, those who play the infinite with wisdom, adaptability, and courage.

About the Author

Maman Ibrahim is a seasoned cybersecurity and digital risk executive with over 20 years of experience advising global organizations on navigating complex risk landscapes, strengthening cyber governance, and unlocking business value through the strategic use of emerging technologies. His career spans leadership roles at industry giants such as GSK, Michelin, and Haleon, where he led global audit, cyber assurance, and digital risk functions across diverse sectors including pharmaceuticals, manufacturing, and business services.

A passionate advocate for responsible innovation, Maman Ibrahim focuses on helping senior technology and business leaders harness the power of AI to enhance resilience, drive performance, and align cyber strategy with enterprise goals. He is a trusted board advisor, public speaker, and mentor to rising cybersecurity leaders.

Maman Ibrahim currently serves as founder of Ginkgo Resilience LTD and principal partner at EugeneZonda Cybersecurity Consultants, where he supports clients in cyber resilience and operationalizing AI-driven risk management with integrity, foresight, and impact.

Email: maman@mamanibrahim.com

Website: https://mamanibrahim.com

LinkedIn: https://www.linkedin.com/in/mamane

CHAPTER 13

HUMANS + MACHINES: NAVIGATING THE INFLECTION POINT OF AI

By Anupriya Jain
Leading Generative AI Transformations for Fortune 50
Seattle, Washington

Change is the law of life and those who look only to the past or present are certain to miss the future.

—John F. Kennedy

Today, as AI and technological innovations reshape our world at unprecedented speed, this wisdom resonates more powerfully than ever. While each wave of advancement from generative AI to autonomous systems brings its share of disruption, embracing these changes isn't just an option, it's imperative for our collective progress toward a more capable and connected future.

I belong to a generation that has lived through some of the most transformative technological shifts in human history. I grew up in India during a time when technology was just beginning to weave itself into the fabric of everyday life. Black-and-white televisions had already made their mark by the time I was born, but I still remember the excitement when a color television finally arrived in our living room. It instantly became the envy of the neighborhood, and watching those vivid images felt like the world had suddenly come alive.

Then came the internet in the late 1990s—a moment that didn't just bring the world closer, it made it instantly accessible. As I stepped from the sheltered halls of high school into the vast, electrifying wilderness of college life, technology began leaping from desktops into our hands. I saw mobile phones evolve from basic devices that let us make calls and play pixelated snake games into smartphones—sleek, powerful machines that replaced calculators, cameras, music players, and much more. Each shift brought with it a deeper realization that technology doesn't just change how we live, it changes who we are. It transforms our ambitions, the way we work, how we interact, and even what we value.

Later, when I moved to the United States, I found myself at the center of yet another wave of innovation. I wasn't just witnessing the change, I was helping shape it. This journey, from growing up in a rapidly transforming India to working in the heart of the global tech industry, shaped how I see artificial intelligence today: not as a sudden disruption, but as the next chapter in a long, ongoing story of human ingenuity and adaptation.

I became the leader responsible for bringing AI and automation into the operational core of one of the world's largest financial institutions—Citigroup. I led initiatives to automate core back-office operations for the bank's credit card division such as detection and prevention of account fraud, claims adjustment, invoice processing—functions historically handled by a large workforce of customer service agents. But from the very beginning, I saw AI not as a substitute for people, but as a partner, one that could remove the mundane work and free up human energy for higher-order tasks. This philosophy came to life in an unexpected and powerful way during the COVID-19 pandemic. With global workforces constrained, the automation systems we had built weren't just helpful, they became critical infrastructure. They allowed essential operations to continue, helped process the US government paycheck protection program's loans to impacted small businesses nationwide, detected fraud in stimulus checks disbursed by the US Department of Treasury, and ensured that the digital heartbeat of the bank didn't stop. In those moments, I saw AI at its best: not just efficient, but humane—*not just smart, but in service of something greater.*

And then, just as the world was catching its breath, came another seismic shift, this time in the very nature of intelligence itself. With the emergence of generative AI, machines began doing what once felt exclusively human: creating. These systems could write coherent essays, draft legal documents, generate code, create realistic images, and even hold multi-turn conversations. They weren't just automating tasks but expanding the boundaries of thought, communication, and creativity. It was during this pivotal moment that I joined Amazon to lead a groundbreaking program for development of one of the first large-scale industry implementations of conversational AI through Alexa. The goal was no longer to just respond to commands, but to create an assistant that could understand context, hold natural conversations, and assist users in real, intuitive ways, blending language, reasoning, and personalization to make technology feel less like a tool and more like a collaborator.

This chapter is an attempt to frame that experience within the broader arc of human history and to explore how we arrived at this moment, what we can learn from the past, and why this isn't just another wave of innovation. It is a civilizational inflection point, one that mirrors the most profound transformation humanity has ever undergone, the Industrial Revolution.

The First Age of Automation: Industrial Revolution

What were the most defining moments in human history? Was it the rise and fall of mighty empires, the great conquests, the discovery of new continents, or the reigns of kings and queens? Yet, if we measure history by how it transformed the lives of ordinary people, a fascinating pattern emerges. For millennia, the average person's way of life remained remarkably unchanged—a farmer in 1500 lived much like a farmer in 500 BCE, with similar GDP per capita century after century.

Then came the watershed moment with James Watt's steam engine invention in the late 1700s. This single innovation sparked the Industrial Revolution, setting in motion a technological transformation that would revolutionize human existence. Today, we enjoy a GDP per capita 50 times greater than that of pre-industrial times. This stark contrast reveals a profound truth that while empires and kingdoms shaped political boundaries, it was a technological innovation that truly revolutionized the human experience, lifting billions from subsistence living to unprecedented prosperity.

Beginning in the mid-18th century in Britain and spreading across Europe and America, the Industrial Revolution marked the move from handmade to machine-made, from cottage industries to factories, and from rural agrarianism to urban capitalism. At its core was automation and the mechanization of labor. The steam engine replaced horses. The loom replaced weavers. Assembly lines replaced artisan workshops. Productivity skyrocketed. For the first time in human history, GDP began to grow exponentially, driven not by population growth or land expansion, but by technological leverage.

This explosion in wealth came with turbulence. Millions of workers were displaced. The Luddites famously rebelled against the machines that replaced their livelihoods. The benefits of progress were not evenly distributed. Yet over time, new industries emerged. Entire classes of jobs were created, from engineers to factory managers, mechanics to marketers. Public education expanded to meet the need for literate, trainable workers. Labor laws, unions, and social reforms helped rebalance the scales. The workforce evolved and, in many ways, came out stronger. The lesson? Disruption is inevitable. Degradation is optional.

Today's Inflection Point: The AI Revolution

Now, we are at a similar moment, arguably even more profound. If the Industrial Revolution gave us mechanical muscle, the AI revolution gives us artificial cognition. We are witnessing the emergence of machines that can not only do, but learn, adapt, and decide. This shift is happening across every layer of society. AI has a far-reaching influence, integrating essential capabilities like classification, prediction, and diagnosis into every industry. It is diagnosing diseases with high accuracy, optimizing logistics and supply chains, trading strategies, assisting teachers, scientists, and designers in real time, and even writing poetry.

What makes this revolution unique is not just the power of AI but its speed and scale. We're not seeing a linear progression, but rather we're at the base of a hockey stick curve. The "hockey stick curve" is a metaphor for exponential growth. At first, progress is slow and barely perceptible. Then, seemingly overnight, it shoots upward. That's where revolutions become irreversible. In AI, we are still in the early stages of this curve. Much of the foundational work like model training, infrastructure scaling, ethical frameworks, and data integration is happening behind the scenes.

To many, it may seem like AI is overhyped or not yet delivering economic miracles. But history tells us that all major technologies start this way. The steam engine took decades before it reshaped transportation and manufacturing. Electricity was known for 50 years before it was widely adopted in homes and factories. The internet was considered a niche technology until broadband and smartphones tipped the scale. AI is following the same pattern. We are building the foundation now, and once the inflection point is crossed, the effects will be rapid and nonlinear. Productivity will surge. Entire industries will be reshaped. Just-in-time workflows will evolve into real-time intelligence ecosystems. The economic impact won't just grow, it will accelerate.

Like every revolution before, the AI era brings a central question: What happens to the people? In the Industrial era, manual laborers were displaced by machines. In today's world, some jobs face the same challenge. Take the example of a customer service representative. Traditionally, much of their day was spent answering routine inquiries such as checking order statuses, resetting passwords, and processing returns. With AI, especially large language models, many of these interactions can now be handled instantly, around the clock, and in multiple languages. But this doesn't eliminate the human role, it elevates it. As AI handles the high-volume, low-complexity work, representatives are freed to focus on higher-value tasks of resolving emotionally charged or complex cases, managing exceptions, and turning service issues into customer loyalty moments.

Meanwhile, new roles emerge. Some agents become conversation designers, training AI systems to speak in the brand's voice. Others become supervisors, reviewing AI interactions, handling escalations, and ensuring that the AI behaves ethically and empathetically. And with the rise of agentic AI, systems that can autonomously act, delegate subtasks to other AI agents, and interact with internal tools, this potential expands even further. An agentic customer support assistant could not only answer a query but also initiate a refund, rebook a service, or loop in a human when it hits a confidence threshold. This redefines the human's role again from a responder to an orchestrator, overseeing a dynamic interplay of AI agents, business logic, and human empathy.

Next let's consider healthcare. Doctors and nurses today often spend as much time on documentation as they do with patients. Generative AI is already starting to change that and acting as a medical scribe that listens during consultations and automatically drafts clinical notes. But again, it's not about replacement, it's about rebalancing. Healthcare professionals still validate and edit the output, apply clinical judgment, and ensure the narrative is accurate and empathetic. Meanwhile, administrative staff shift into roles like AI system reviewers, data privacy monitors, and care workflow designers ensuring that these tools align with medical ethics and regulations like HIPAA. As agentic AI matures, it could go beyond note-taking to coordinating care, scheduling follow-ups, sending reminders, summarizing lab results, or flagging anomalies across systems. In such a future, healthcare workers collaborate with AI not just as tools, but as partners tasked with maintaining trust, ensuring safety, and focusing on the human connection at the heart of care.

These examples show us that the future of work isn't about humans versus machines. It's about humans working with machines, and we need to rethink how to prepare people for that future. We need learning to be ongoing and accessible throughout life, not just through one-time college degrees. People will need to regularly update their skills, especially when it comes to working alongside AI and knowing how to use it, question it, guide it, and improve it. Our workforce training programs also need to shift focused on building AI literacy. Understanding what AI can and can't do, how to supervise it, how to correct it, and how to work with more advanced systems like agentic AI, which can take actions, make decisions, and even manage tasks on their own. And we'll need new rules too. Policies that define who's responsible when AI systems act, how to keep them safe and fair, and how to make sure they support people and not replace them. Governments, educators, and businesses will all need to work together to build this support system just like they did in the past in the Industrial Revolution.

Final Thoughts: From Steam to Semiconductors

The Industrial Revolution didn't just change how we built things but changed how we thought about work, value, and progress. The AI revolution will do the same. But unlike our ancestors, we have history as a guide. It shows that with every major shift, new kinds of work emerge. The real question is: Will we be ready for it?

If we're not, we risk repeating the mistakes of the past where progress left too many people behind. But if we plan ahead, we have a rare chance to do better this time and to shape a future where technology lifts everyone up.

And that's what this chapter is really about: understanding how we got here, what's changing now, and why this moment is different. This isn't just another wave of innovation. It's a turning point like the Industrial Revolution. And if we get it right, we won't just create smarter machines—we'll create a wiser, more inclusive, more resilient world.

About the Author

Anupriya Jain is a seasoned leader in the field of artificial intelligence, with over two decades of global experience driving innovation at the intersection of technology, business, and society. She is a University Gold medalist in Electronics and Instrumentation Engineering from India and has further strengthened her strategic acumen through an executive program in digital business strategy from MIT Sloan School of Management. Anupriya has led transformative AI initiatives across cloud, machine learning, and robotic process automation, with a focus on scaling intelligent systems to optimize business operations and combat fraud. Her pioneering AI solution enabled the disbursement of $500 million in COVID-19 pandemic relief to US small businesses and prevented widespread stimulus fraud—earning her Citi's prestigious Leaders in Excellence Award. Most recently, she spearheaded the launch of the next-generation Alexa powered by generative AI to deliver personalized and proactive conversational experiences. In addition to her professional work, Anupriya actively volunteers in her local community, promoting STEM education and encouraging young minds to explore emerging technologies.

LinkedIn: https://www.linkedin.com/in/anupriya-jain/

CHAPTER 14

THE RACE TO ROI: THE REAL KEY TO WINNING WITH AI

By Lydia James
Transformation Coach, Speaker, CX Thought Leader
Houston, Texas

Build a strong foundation and you can reach even the most unthinkable heights.

—M.J. Moores

Artificial intelligence has rapidly evolved from an emerging technology buzzword to a critical boardroom priority. Organizations across industries are investing heavily in AI with promises of automation, precision, and predictive insights that could revolutionize their operations. Yet beneath this enthusiastic adoption lies a concerning reality: The majority of these initiatives fail to deliver meaningful return on investment.

Why do so many AI projects falter? The answer isn't in the technology itself—it's in the transformation approach. After two decades leading transformation initiatives and leaders across operations, sales, marketing, and technology, I've observed a consistent pattern: Technology is only as powerful as its implementation strategy. The difference between AI success and failure rarely comes down to the sophistication of algorithms, but rather how effectively organizations bridge the gap between technological capability and business application. This expanded perspective unpacks how to overcome common AI pitfalls, amplify success factors, and fast-track ROI through strategic planning, organizational structure, and human-centered execution.

The Messy Middle: Where AI Initiatives Get Lost

Every organization has business problems worth solving—and AI providers offer increasingly powerful capabilities. But between these two realities exists what I call "the messy middle": a zone characterized by confusion, misalignment, scope creep, and execution gaps. In this uncertain territory, critical questions often go unasked:

- Are we solving the right problem, or are we captivated by a solution looking for a problem?
- Are we investing in capabilities we genuinely need—or simply those that are available?
- Do we require the full spectrum of AI capabilities or just specific components?
- Where should we begin to ensure early success and organizational buy-in?
- How will this initiative align with our existing systems, processes, and culture?

This gray zone is precisely where ROI is either secured or squandered. Recent research illuminates the scope of the challenge:

- According to Gartner, up to 85% of AI projects fail to deliver on their intended promises.

- MIT Sloan and BCG found that a mere 11% of companies report significant financial benefits from their AI investments.
- Over 40% of AI initiatives never scale beyond the proof of concept or pilot phase, resulting in sunk costs and stalled momentum.

The disconnect between AI's theoretical potential and practical business value represents billions in wasted investment and countless missed opportunities for genuine organizational advancement.

Why AI Projects Fail (And What That Costs You)

Let's address the uncomfortable truth—AI typically doesn't fail because the technology is fundamentally flawed. It fails because organizations don't set these initiatives up for success from the beginning.

Four Common Reasons AI Initiatives Go Sideways

1. Unclear Problem Definition (Scope)

    - Ambiguity around what problem the AI solution is truly solving
    - Insufficient documentation of current processes and their execution
    - Inconsistency in how processes are performed across the organization
    - Limited understanding of how the problem impacts customers, revenue streams, or compliance requirements
    - Scope creep as additional "nice-to-haves" get incorporated without strategic evaluation

2. Missing Stakeholders (People)

- Lack of clear ownership for ensuring successful outcomes
- Key decision-makers excluded from early planning phases
- Insufficient cross-functional buy-in, especially from eventual end users
- Silent resistance from stakeholders who feel threatened but aren't given voice
- Overlooking cultural implications and change management requirements

3. Inadequate Resources (Budget)

   - Underestimating implementation costs beyond the technology purchase
   - Insufficient time allocation for testing, training, and adaptation
   - Limited human resources dedicated to implementation and optimization
   - Failure to account for opportunity costs of competing initiatives
   - Missing budget for ongoing maintenance, upgrades, and evolution

4. Misaligned Technology (Product)

   - Selecting AI solutions based on hype rather than strategic fit
   - Applying sophisticated technology to poorly defined problems
   - Pursuing full automation when augmentation would deliver better results
   - Choosing point solutions that solve isolated problems without integration potential

- Implementing technologies that don't align with existing technical infrastructure

When AI enters this equation, the stakes become substantially higher. Failed projects aren't merely costly—they can inflict lasting organizational damage:

- Potential exposure of sensitive customer or operational data
- Erosion of trust among teams, executives, and external stakeholders
- Direct financial losses through wasted investment, delayed revenue, and diminished customer loyalty
- Significant setbacks in broader digital transformation momentum
- Development of long-term organizational resistance to future innovation initiatives

These consequences extend far beyond the immediate project failure, creating a ripple effect that impacts organizational culture, market position, and competitive agility.

Learning from Success and Failure: Contrasting Case Studies

Case Study 1: A CRM Call Summary Project Gone Sideways

- *Project goal*: Automate post-call summaries for customer service agents and seamlessly input them into the CRM system.
- *Expected ROI*:

- Save approximately two minutes per call for each agent
- Achieve $1 million in cost savings through improved efficiency
- Enhance data quality and consistency in customer records

- *What actually happened:*

    - *Poor process definition*: While the AI successfully summarized calls, agents still needed to review, modify, and manually transfer the output into the CRM—actually adding time to their workflow rather than reducing it.
    - *Inadequate data strategy*: Critical integration points that could have trained the AI faster and enhanced accuracy were overlooked during implementation.
    - *Siloed implementation*: Multiple teams ran separate AI pilots with no coordination, shared learning, or data integration strategy.
    - *Stakeholder exclusion*: CRM users weren't meaningfully consulted on how they would use the new output or how it should be formatted for their needs.
    - *Missing support infrastructure:* A third-party consultant team built and deployed the solution with no ongoing support or evolution plan once they departed.
    - *Cultural disruption:* Lack of transparency with frontline teams led agents to feel they were being monitored without consent, creating resistance to participate or provide constructive feedback.

- *Outcome*:

    - Unbudgeted cost increases across implementation and adaptation
    - Significant delays in realizing any operational benefits

  - Frustrated and disengaged agents whose workflow was disrupted rather than enhanced
  - Leadership questioning the fundamental value of AI investments across the organization
  - Damaged credibility for the project sponsors and technology advocates

Case Study 2: A Quality Analytics Automation Project Win

- *Project goal*: Automate the quality assurance process to evaluate 100% of agent calls rather than a small sample.
- *Expected ROI*:
  - Save ten minutes per evaluated call for QA specialists
  - Expand evaluation coverage from sample-based to 100% of customer interactions
  - Achieve $1 million in cost savings through efficiency and quality improvements
- *What actually happened*:
  - *Process refinement first*: The team standardized the quality evaluation process across all departments before introducing automation, creating a consistent foundation.
  - *Comprehensive data strategy*: Key performance indicators were clearly defined, and a robust data repository was established to support AI integration.
  - *Coordinated implementation*: One core team led the effort enterprise-wide, providing regular progress updates and ensuring a consistent approach.

- *Engaged stakeholders*: Stakeholder roles and responsibilities were explicitly outlined from the beginning, with clear expectations and decision rights.
- *Sustainable support model*: The project was led by an internal team partnering with third-party providers, including process owners to ensure ongoing continuity.
- *Cultural enhancement*: Trust increased as agents gained confidence in being evaluated on 100% of their interactions rather than potentially unrepresentative samples. Process owners who had been involved from conception became natural champions for the change.
- *Feedback integration*: A structured feedback process allowed continuous improvement and strategic pivots when necessary.

• *Outcome*:

- Achieved $3 million in cost avoidance, triple the projected savings
- Demonstrated a 3% increase in quality scores across 2,500 agents
- Established a streamlined, consistent QA process across the enterprise
- Advanced from binary pass/fail quality assessments to a more sophisticated competency-based system
- Improved team morale by removing perceived human bias from the auditing process
- Created a template for future AI implementation success

These contrasting cases illustrate that the determining factor in AI success isn't the sophistication of the technology but rather the thoroughness of implementation strategy and execution.

The Winning Formula for AI ROI

Achieving meaningful ROI from AI isn't a matter of chance—it's the result of disciplined leadership, thoughtful structure, and excellence in execution. Here's your comprehensive blueprint for success:

Core Principles

1. *AI must serve business goals, not vice versa.*

Every AI initiative must link directly to tangible business outcomes: cost reduction, revenue growth, improved customer experience, enhanced employee productivity, or competitive differentiation. Technology without clear purpose becomes an expensive distraction.

2. *Leaders must drive AI adoption.*

AI isn't merely an IT project; it's a business transformation initiative requiring cross-functional ownership. Leadership accountability must be crystal clear, with collaboration spanning from C-suite executives to frontline employees. Transparency should be a cornerstone of your change management approach.

3. *Build for augmentation, not elimination.*

AI delivers its greatest value when enhancing human capabilities rather than replacing them. Focus on elevating your team with tools that support more intelligent work, deeper insights, and higher-value activities—not simply reducing headcount. Understanding the full spectrum of effort across teams ensures the delivered solution genuinely addresses pain points.

4. *Data is the foundational fuel.*

No AI project can thrive on poor-quality data. Invest in creating clean, structured, accessible data repositories from the beginning. Develop a clear understanding of how data flows through your organization and how it can provide greater insights for improved business intelligence.

5. *Think lifecycle, not launch.*

AI implementation isn't a one-time event but an ongoing journey. Each experiment should test expected outcomes and incorporate refinements. Sustainable success comes from continuous improvement, real-time feedback mechanisms, and dedicated support structures that evolve with your business needs.

The ROI Blueprint: Eight Steps to AI Success

1. *Start with the right process.*

Choose a well-defined, measurable process with manageable risk for your initial AI implementation. This provides your team space to experiment, learn, and adapt without risking enterprise-wide disruption. Ensure your selected process is:

- Thoroughly documented with clear inputs and outputs
- Consistently repeatable and objectively measurable
- Owned by a committed team willing to champion transformation
- Sufficiently important to demonstrate value, yet contained enough to manage risk

A solid process foundation creates the conditions for smarter, faster AI adoption and establishes a template for future initiatives.

2. *Assemble the right people.*

Identify your power team from the beginning:

- *Executive sponsor*: a senior leader with organizational influence and resource authority

- *Business owner*: an operational leader directly accountable for outcomes
- *Stakeholders*: cross-functional experts representing every touchpoint in the process
- *End users*: the people who will ultimately work with the AI solution daily

Early alignment prevents resistance, ensures adoption, and provides diverse perspectives that strengthen implementation. Cross-functional collaboration is your greatest risk mitigator and innovation accelerator.

3. *Build a solid data strategy.*

AI solutions are only as good as the data that powers them. Clean, structured, reliable data is non-negotiable. Ask critical questions:

- Is the data complete, appropriately tagged, and consistent across sources?
- Can it be easily accessed and processed by AI tools?
- Are exceptions and edge cases represented in the dataset?
- Who owns the data quality and how will it be maintained?

Remember: garbage in = garbage out. A weak data foundation will inevitably derail even the most promising initiative.

4. *Prioritize data compliance.*

AI implementation amplifies data risks, making compliance not merely optional but mission-critical:

- Thoroughly understand relevant data privacy regulations (GDPR, HIPAA, CCPA, etc.).
- Conduct comprehensive data risk assessments before implementation begins.

- Establish strict controls for access to sensitive data with robust audit trails.
- Partner closely with legal and compliance teams to validate use cases.
- Ensure AI outputs are explainable and decisions can be traced to their origin.

Compliance isn't a bottleneck—it's a brand protector and trust builder that safeguards your organization's reputation and customer relationships.

5. *Set clear ground rules.*

Define critical parameters upfront:

- Success metrics (specific KPIs, cost savings targets, experience metrics)
- Decision rights (who can make which decisions at each implementation stage)
- Escalation pathways for addressing challenges or conflicts
- Project rituals (meeting cadence, review processes, reporting structure)

Approach AI implementation with the mindset of a transformation leader, not just a technology buyer.

6. *Get crystal clear on scope.*

Ambiguous problem definitions inevitably produce ambiguous results. Clarify:

- The specific business challenge being addressed
- Detailed mapping of current process states and pain points
- Precisely defined target state with measurable KPIs
- All integration touchpoints with existing systems

- A phase-based implementation approach with clear milestones

7. *Plan for resistance.*

Change inevitably triggers emotional responses—address these proactively:

- Communicate the "why" behind the AI initiative with transparency and clarity.
- Meaningfully involve those whose work will be impacted in the design process.
- Offer comprehensive reskilling and transition support for affected team members.
- Create psychological safety for expressing concerns and contributing ideas.

Remember: People don't inherently resist change; they resist being left behind or overlooked in the process of change.

8. *Build for the long game.*

AI isn't a static tool but a living system requiring ongoing attention:

- Develop structured frameworks for continuous AI training and refinement.
- Establish formal feedback loops that capture insights from users and outcomes.
- Schedule regular data refreshes to maintain accuracy and relevance.
- Create a thoughtful roadmap for scaling successful applications across functions.
- Invest in knowledge transfer to develop internal AI expertise.

From Hype to High-Performance: Making AI Work for You

When companies implement AI with purpose, precision, and people at the center, transformational results emerge:

- Teams reclaim valuable time and make more informed decisions.
- Customers receive more personalized, consistent, and responsive service.
- Executives see measurable ROI that justifies further investment.
- Organizations gain deeper, actionable insights that drive competitive advantage.
- Operations achieve unprecedented consistency and speed at scale.
- Innovation accelerates as teams focus on high-value work rather than repetitive tasks.

These outcomes become possible only when AI serves your business strategy—not when your business contorts itself to accommodate trendy technology.

Final Perspective: Execution Trumps Technology

Technology itself doesn't transform organizations—execution does. AI holds immense potential, but power without direction is, at best, wasteful and, at worst, dangerous. Your implementation strategy is the critical differentiator between AI success and failure. You don't need more sophisticated AI. You need smarter implementation.

When you lead with people, process, data, and compliance—rather than chasing capabilities—you position your organization to capture genuine value and establish competitive advantage. The organizations that will win in the AI era aren't necessarily those with the largest technology budgets or the most advanced algorithms.

The winners will be those who excel at translating AI's theoretical potential into practical business value through disciplined execution and human-centered design.

About the Author

Lydia James is the founder and chief transformation officer of The Pivot Solutions Group, partnering with business leaders to fast-forward project success and capture return on investment while leading courageously and authentically through disruptive times. With 20-plus years of experience across military, Department of Defense, retail, and automotive sectors, Lydia is recognized for her ability to drive large-scale enterprise transformation, optimize customer experience (CX), and implement innovative service strategies to improve client loyalty and employee engagement.

As a transformational coach, a speaker, and a CX thought leader, she has been invited to speak at industry conferences, corporate leadership summits, and executive panels, sharing practical, data-driven strategies on business growth, transformation, CX excellence, and the impact of AI-driven automation and predictive analytics in shaping the future of customer experience and operational efficiency. Lydia serves on the advisory board for Customer Contact Week (CCW), a leading organization dedicated to transforming the world of CX. In this role, she provides strategic guidance on AI innovation, automation, and service strategy transformation, helping organizations redefine customer engagement through AI-driven solutions and predictive analytics. She plays a key role in shaping the future of intelligent customer interactions, next-generation self-service models, and the ethical application of AI in CX.

Lydia's has been recognized by Customer Connect Expo for her exceptional leadership in CX strategy, vision and execution in driving transformational changes within organizations, and inspiring progress across the customer experience ecosystem. She was featured in Women Worth Watching, CX Network, and CCWomen Stronger Together magazine, where she shares insights on business transformation, leadership agility, and customer experience innovation. She was also named one of CX Network's "20 Leaders to Watch" for her visionary approach to revolutionizing CX through AI-driven customer insights and data analytics. As Lydia says, "Transformation isn't just about strategy—it's about leadership, vision, and execution. When leaders evolve, businesses thrive."

Email: Connect@thelydiajames.com

LinkedIn: https://www.linkedin.com/in/lydiavjames/

CHAPTER 15

EMPOWERING THROUGH AI: CYBERSECURITY, CLOUD, AND THE HUMAN SIDE OF TECH

By Bridget Kovacs
Senior IT Manager, Cybersecurity Expert
Sarasota, Florida

You can't stop the waves, but you can learn to surf.

—Jon Kabat-Zinn

The Shift We're All Living Through

Whether we realize it or not, we're living through one of the most significant technological shifts in human history. Artificial intelligence is no longer some far-off concept tucked away in research labs—it's already here, shaping the way we work, learn, create, and connect. And for a lot of people, that feels overwhelming.

I get it. I've worked in tech for nearly two decades—leading teams, building infrastructure, defending systems from cyber threats—and even I've had moments where the pace of change made me pause. But here's what I know for sure: AI doesn't have to be something that happens *to* us. It can be something we shape, something we partner with, and something we use to create a more secure, creative, and connected future.

This chapter is about empowerment over fear. It's about helping people see that AI isn't just for programmers or data scientists. It's for *everyone*—from busy parents to small business owners to seasoned professionals who are learning to pivot in a fast-changing world. It's also about the human side of all this tech—the mindset, the skills, and the values that will help us thrive in a world that's increasingly digital, yet still deeply human.

Cybersecurity in the Age of AI

As AI becomes more advanced, so do the threats. Gone are the days when cybersecurity was just about installing antivirus software or changing your password every six months. Today's risks are faster, more sophisticated, and often invisible to the untrained eye. AI has become a double-edged sword—helping us detect threats in real time, but also powering the tools that hackers use to breach systems more creatively than ever before.

But here's the part that often gets missed: Cybersecurity is no longer just an "IT thing." It's an *everyone thing*. If you have an email account, a phone, a social media presence, or you run a business, then cybersecurity is part of your life, whether you like it or not.

Fortunately, you don't need to be a security expert to protect yourself. Here are a few simple ways to stay safer in the AI era:

- Use multi-factor authentication wherever possible.
- Don't reuse passwords across accounts—password managers can help.

- Be mindful of phishing—especially AI-generated scams that look eerily legitimate.
- Regularly back up your important files to the cloud or an external drive.
- Keep your software updated—yes, those annoying updates actually matter.

Most importantly, stay curious. Ask questions. If something feels off, trust your gut and investigate. AI can be an incredible ally in cybersecurity—scanning logs, predicting attacks, and plugging holes before humans even notice. But that only works when we're part of the process. When we stay informed, we stay empowered.

The Cloud and Real-Life Use Cases

Let's talk about "the cloud." It sounds fuzzy and abstract, but the truth is, you're probably using cloud-based AI tools every day without even thinking about it.

- When your phone auto-sorts your photos? That's AI in the cloud.
- When your email filters spam or your GPS recalculates mid-drive? Cloud and AI, working together.
- When your business uses Google Docs, Zoom, or customer service chatbots? Yep—more of the same.

The cloud isn't just about storage—it's about accessibility, speed, and scalability. And when you layer AI on top of it, the possibilities multiply. Small businesses can now automate scheduling, customer interactions, and even marketing campaigns without needing a full IT department. Individuals can use tools like ChatGPT, Grammarly, or Notion AI to level up their writing, planning, and learning—often for free or at a very low cost.

For people who didn't grow up with this technology, it can feel like a lot. That's why I always suggest starting small. Pick one tool.

Explore one platform. Ask someone you trust to walk you through it. You don't need to become an expert overnight. You just need to stay open and willing to try.

The Human Side: Mindset, Adaptability, and Purpose

This is the part people often overlook. We talk about tech like it's the whole story, but it's only half. The other half—maybe the more important half—is us. In a world filled with automation, the most valuable skills are the ones that only humans can bring to the table: emotional intelligence, critical thinking, curiosity, creativity, and adaptability. These aren't just nice-to-haves. They're *power skills.*

I've seen firsthand how people can thrive when they embrace change instead of fearing it. After years of working in government and corporate IT, I took a career break to care for my son. It was one of the most beautiful, disorienting, and transformative seasons of my life. I used that time to study, explore, and eventually shift my focus toward AI, cloud, and helping others navigate change. I didn't have a master plan—just a willingness to learn, to be wrong, and to keep going.

That's what it takes now. Not perfection. Not a ten-step blueprint. Just courage, curiosity, and the belief that you still matter in this new world—maybe more than ever.

Getting Started: One Powerful Tool to Begin With

If you're just starting out with AI, the best place to begin is with the free version of ChatGPT. It's like having a smart assistant who's always ready to help—whether you're figuring out dinner with random ingredients, need a substitute in a recipe, want help planning a project, or even need a spreadsheet built from scratch. The more I use ChatGPT, the more I realize just how versatile it is. I now use it every single day—for everything from quick information lookups to writing support, creative ideas, and even image generation. I've explored a range of AI tools—including Gemini, CoPilot, Midjourney, Canva, and Runway—covering everything from writing assistance to design and automation. While each has its strengths, ChatGPT continues to be the most practical and consistently useful tool in my daily life. My advice: Anytime you're about to Google something, try asking ChatGPT instead. You might be surprised at how much easier—and more personalized—the answer is. And if a task feels too big, just ask AI for help.

A useful tip on very large projects or tasks: Break it down into smaller parts. That's when AI really shines.

Conclusion: Empowered, Not Replaced

We can't slow down the pace of change, but we can choose how we respond to it. AI is not here to replace you. It's here to challenge you to evolve—to think differently, to work smarter, and to bring more humanity into a world that desperately needs it.

The truth is, AI is only as powerful as the people behind it. That means you still matter. Your voice, your choices, your values—all of it shapes the kind of future we create. So, let's not be afraid of what's coming. Let's get curious. Let's get equipped. And let's move forward—together.

About the Author

Bridget Kovacs is a seasoned IT leader with over 17 years of experience in cybersecurity, cloud infrastructure, and enterprise systems. She has led major tech initiatives for government and private organizations and is passionate about making technology feel more human and less intimidating. Bridget lives in Sarasota, Florida, with her husband and son, where she enjoys exploring parks, spotting wildlife, and riding around the neighborhood in the family's golf cart. She also enjoys visiting all the beautiful beaches in the area, boating, and outdoor adventures with friends and family. A lifelong learner, Bridget believes that curiosity is one of the most important skills we can nurture in the AI era.

LinkedIn: www.linkedin.com/in/bridget-kovacs

CHAPTER 16

HOW TO PREVAIL IN THE AGE OF AI

By Matt Kurleto
Serial Entrepreneur and AI Strategy Consultant
Gdansk, Poland

We can design AI so that humans are always in the loop. We can design AI so that AI is empowering people, not displacing people.

—Satya Nadella, CEO of Microsoft

Let's go back to 1950, Manchester. In a small office, Alan Turing posed a question that would ripple through time: "Could a machine ever convince a human it was thinking?" His "imitation game" lingered as a thought experiment for decades—fascinating, but far from reality.

Fast forward to 2022. Without fanfare, ChatGPT quietly entered our browsers. But what followed was anything but quiet. Suddenly, millions were chatting with a machine that could reason, empathize, even create. The Turing test wasn't just theoretical—it had gone mainstream. Lawyers, marketers, doctors, and poets all found new creative partners in AI. It was no longer science fiction—it was a new interface for human imagination.

Markets surged. FOMO spread. And, as with every revolution, hype buzzed louder than understanding. But beneath it, a familiar rhythm played out: wonder, speculation, and then—slowly—transformation. Just like the early internet, AI felt raw and full of untapped potential.

Today, AI sits at a crossroads. It's powerful, but rough around the edges. Some fear it will replace us. But history shows technology amplifies more than it erases. The printing press empowered storytellers. The internet expanded conversations. AI, too, is here to multiply—not mute—human creativity.

At Neoteric, we believe AI is not one thing, but many. Like any team, it's made up of personalities. We see three key AI archetypes: *monkeys*, who tirelessly automate tasks; *experts*, who analyze data and optimize decisions; and *interns*, the creative but sometimes erratic generative tools. To work with AI, you must understand these AI archetypes—not as threats, but as teammates.

The future won't be human-only or machine-only. It will be a hybrid—collaborative, dynamic, and deeply human at its core. The age of AI isn't the end of the story. It's where things get interesting.

The AI Monkey: Relentless Repetition, Flawless Execution

Our story begins on the digital assembly line, where the AI monkey reigns supreme. Imagine a tireless worker, one who never takes a coffee break, never gets distracted, and never complains about the monotony of the task at hand. This is the realm of computer vision models that scan thousands of images for defects, of classification systems tagging mountains of data, of fraud detectors sifting through endless streams of transactions. The AI monkey doesn't innovate or strategize—it simply executes, again and again, with a precision that borders on the uncanny.

In the early days of automation, these were the first AI agents to join our teams. They brought speed, scale, and objectivity. But they also taught us a valuable lesson: Don't ask your experts to do a monkey's job, and don't expect your monkey to write poetry. The monkey is a master of repetition, not of reinvention.

The AI Expert: Pattern Seeker, Insight Giver

As our digital workforce matured, a new archetype emerged: the AI expert. If the monkey is the assembly line worker, the expert is the seasoned analyst who can sift through mountains of data and spot patterns invisible to the naked eye. Fueled by deep learning and predictive algorithms, these agents became our advisors and forecasters. They tell us which movie we might enjoy, which product to recommend to a customer, when a machine is likely to fail, or how to optimize a supply chain.

The expert doesn't create new ideas out of thin air, but it excels at making sense of the chaos, surfacing insights, and guiding decisions. It is the quiet force behind recommendation engines, predictive maintenance systems, and personalized marketing campaigns. The AI expert is not a dreamer, but a relentless optimizer—always searching for the next marginal gain.

The AI Intern: Creative, Unpredictable, and Full of Potential

Then, almost overnight, a new face appeared in the digital office: the AI intern. This is the generative AI we've all heard about—the one that drafts emails, designs logos, brainstorms campaign ideas, and even negotiates deals. The intern is eager, creative, and sometimes astonishingly insightful. But, like any intern, it's also prone to the occasional blunder, misunderstanding, or burst of overconfidence.

The AI intern is the wildcard. It can be brilliant one moment and baffling the next. It needs guidance, feedback, and a steady hand to channel its energy productively. But when managed well, the intern becomes a formidable copilot, helping us explore new ideas, accelerate creative work, and push the boundaries of what's possible.

A New Kind of Teamwork

What emerges from this cast of characters is a new model of collaboration. AI isn't just a tool; it's a workforce—a digital team with roles as varied as any human organization. The lesson is clear: Success in the age of AI isn't about replacement, but orchestration. It's about learning to delegate, to collaborate, and to lead. The most effective leaders will be those who can recognize the strengths of each archetype, assign the right tasks to the right agent, and foster a culture where human and machine talents amplify each other.

Just as a great manager wouldn't ask a data scientist to sort mail or a junior intern to set company strategy, we shouldn't expect our AI agents to be all things at once. Instead, we must become conductors—tuning, guiding, and harmonizing this new digital orchestra. Only then will we unlock the true potential of AI, not as a replacement for human ingenuity, but as its most powerful collaborator yet.

Let's begin with a scene that's as familiar as it is universal: the early morning desk, a mug of coffee cooling beside a laptop, and a to-do list that seems to multiply by the hour. For years, we've searched for the secret to productivity—new apps, clever hacks, color-coded planners—always hoping to find that elusive edge. But a quiet revolution is underway, and its name is AI.

We must learn to transform ourselves before we can hope to transform our teams or organizations. The journey starts at the individual level, where the right tools can turn a single person into a productivity powerhouse. Think of ChatGPT, Perplexity, and their kin—not as novelties, but as force multipliers, ready to amplify your output if you know how to wield them. Like any tool, their impact depends on the skill and intention of their user.

The Three Levels of AI Delegation: A New Playbook for Personal Mastery

As with any craft, mastery comes in stages. In the realm of AI-powered productivity, these stages reveal themselves as three distinct levels of delegation—each a step up in trust, complexity, and impact.

Level One: Task Delegation—The Art of the Precise Ask

Picture this: You're buried under research papers, or perhaps you need to send out a dozen cold emails before lunch. At this first level, you treat your AI as a diligent assistant. You define the outcome, select your tools, outline your assumptions, and specify the process. The instructions are clear, the boundaries well-drawn. "Summarize this research," you might say. "Draft a cold email to this prospect." "Tag my meeting notes for follow-up." The AI executes these requests with speed and accuracy, freeing you from the tyranny of the mundane.

It's a bit like having a junior staffer who never tires, never misses a detail, and never complains about repetition. The key here is clarity—what you ask is what you get. The more precise your delegation, the better the result.

Level Two: Goal Delegation—Orchestrating Outcomes

As your confidence grows, you begin to see the bigger picture. Instead of handing off isolated tasks, you start to delegate clusters of related activities, all aimed at a defined goal. Here, you're not just asking for a summary or a draft—you're asking your AI to "optimize our onboarding experience" or "improve open rates across channels." Now, the AI becomes a project collaborator, breaking down your request into smaller steps, tracking progress, and measuring success against clear metrics.

At this level, you're orchestrating outcomes, not just actions. You provide the vision and the guardrails, and let the AI handle the heavy lifting of execution, analysis, and iteration. It's a partnership—your strategic intent, the AI's operational muscle.

Level Three: Responsibility Delegation—Entrusting the Ongoing

The final stage of AI integration is full responsibility delegation—trusting your AI to manage entire domains like "weekly trend analysis" or "newsletter creation," with only occasional feedback. The AI adapts, learns, and improves continuously, becoming more like a trusted lieutenant than a passive tool. Your role evolves from taskmaster to mentor, offering guidance while the AI anticipates needs and solves problems independently.

The New Productivity Engine

Top AI users don't just follow trends—they build custom productivity systems. They ask: "What slows me down? What can be automated?" Then, they connect tools and workflows to amplify their unique strengths.

AI isn't about replacing creativity—it's about unlocking it. By automating the repetitive, it frees up space for high-value, strategic work only humans can do. So next time you sit down to work, remember: Productivity isn't about speed—it's about partnership. And your smartest partner may be just a prompt away

Phase Two: Empowering Teams to Work Smarter

Every transformation begins with one person—and grows into a movement. AI adoption is no exception. It's not just a tech upgrade; it's a cultural shift. The real impact doesn't come from software alone, but from how people feel, adapt, and grow together.

Picture a team unsure of what AI means for their future. The leader offers not a list of features, but a promise: "We're here to empower, not replace." In that moment, trust becomes the real foundation.

So how do you create a team that embraces AI? Start with ten simple rules—each grounded in real stories—to guide your people from uncertainty to shared success.

1. *Involve people early.*

Transformation is not something you do to people—it's something you do with them. When a global consulting firm decided to automate parts of their client onboarding, they didn't start with a memo from the C-suite. Instead, they invited frontline staff to a series of workshops, asking: "Where does your work get stuck? What do you wish you could automate?" By surfacing real pain points, they made everyone a co-creator, not just a recipient, of change2.

2. *Over-communicate intentions.*

In times of change, silence breeds suspicion. One startup, rolling out an AI-powered analytics tool, made a point to share weekly updates—not just about what was happening, but why. Leaders explained the vision, the risks, and the expected benefits in every town hall and newsletter. The result? Rumors faded, and curiosity took their place.

3. *Co-create the AI journey.*

When a retail chain introduced AI scheduling, they didn't just drop the new system on employees. Instead, they formed a cross-functional task force—cashiers, managers, IT, and HR—who mapped out how the tool would be piloted, tested, and scaled. This collaborative approach ensured the solution fit real workflows, not just theoretical ones2.

4. *Identify champions.*

Every movement needs its torchbearers. In a mid-sized manufacturer, an enthusiastic process engineer became the "AI champion." She organized lunch-and-learns, demoed new tools, and fielded questions from skeptical colleagues. Her passion and credibility made her the bridge between leadership's vision and the team's day-to-day reality4.

5. *Ask tough questions openly.*

AI isn't magic—it's a tool shaped by the data and decisions behind it. At a financial services firm, leaders invited employees to a "bias audit" session, asking: "Where might our models go wrong? What are we missing?" By surfacing doubts and addressing them head-on, they built a culture of transparency and ethical responsibility5.

6. *Define the "why" behind the change.*

Change for its own sake breeds fatigue. But when a hospital rolled out an AI triage assistant, they anchored every conversation in their mission: "This will help us spend more time with patients, not paperwork." By connecting technology to purpose, they turned skeptics into supporters5.

7. *Visualize the path forward.*

Ambiguity is the enemy of momentum. One logistics company used simple infographics and roadmaps to show how AI would be phased in—what would change, when, and who would be affected. Employees could see themselves in the story, reducing fear of the unknown.

8. *Align AI use with personal growth.*

People embrace change when it helps them grow. A marketing agency offered training and certifications in AI tools, making it clear that learning these skills could lead to promotions and new opportunities. Suddenly, AI wasn't a threat—it was a ticket to advancement.

9. *Share risks and rewards.*

When a team at a SaaS company piloted an AI-powered customer support bot, they agreed upfront to share credit for improvements—and responsibility for any hiccups. When the bot's first week led to a spike in unresolved tickets, no one pointed fingers. Instead, they worked together to fine-tune the system, and celebrated when satisfaction scores rose.

10. *Create structured feedback loops.*

Building an AI-powered team is not about swapping people for algorithms. It's about inviting everyone on the journey, making space for questions, and celebrating progress—together. When trust is the foundation, AI becomes not just a tool, but a teammate. The future belongs to teams that are brave enough to ask, "How can we work smarter?"—and humble enough to answer, "Together."

Phase Three: Strategic Alignment for Maximum Impact

Innovation doesn't start with trendy tools—it starts with the right question. True AI transformation aligns technology with your business's core goals, challenges, and ambitions. That's the purpose of the Neoteric AI innovation funnel: a practical method for turning bold ideas into real, measurable results.

From Idea to Impact: The Journey Through the Funnel

Imagine your organization as a thriving city. Every day, new ideas bubble up from every corner—some inspired by customer pain points, others by bold strategic goals, and still others by the inefficiencies that slow your teams down. The challenge is not a lack of ideas, but knowing which ones deserve your precious time and resources.

Step 1. Identify use cases rooted in strategy and friction.

Start by casting a wide net. Gather ideas from across your company: What's holding back your growth? Where do you see untapped opportunities? Maybe your sales team struggles to qualify leads efficiently, or your operations team faces recurring bottlenecks. Use your company's strategy as the North Star, but don't ignore the market edge or the daily grind—sometimes the smallest friction points reveal the biggest opportunities

Example: A telecom company noticed customer churn was rising. Instead of guessing, they used AI to analyze support interactions and discovered a pattern: Unresolved technical issues were the main driver. This insight became the seed for a targeted AI solution that ultimately reduced churn by over 20%.

Step 2. Prioritize—impact and complexity as your compass.

With a list of potential use cases in hand, it's time to prioritize. Here, the value-versus-effort matrix (or the "quadrant") becomes your map. Plot each idea by its potential impact and implementation complexity:

- *Low-hanging fruits* (easy + high impact): These are your quick wins. Start here—think automated data tagging or AI-powered lead scoring.
- *Game changers* (hard + high impact): These are transformative but require investment and patience—like building a new AI-driven product line.
- *Distractors* (easy + low impact): Avoid these. They're tempting but won't move the needle.

- *Cash burners* (hard + low impact): Kill these fast. Don't let sunk costs or wishful thinking drain your resources.

Example: A SaaS company mapped out dozens of AI ideas. Automated meeting summaries (easy, high impact) went to the top of the list, while a complex, low-impact chatbot idea was shelved before it could eat up resources.

Step 3. Validate—small bets, fast feedback

Don't bet the farm on untested ideas. Instead, run small-scale pilots or "innovation sprints" to test your top priorities. Measure results against clear KPIs. Did the AI reduce manual workload? Did it improve customer satisfaction? If it works, scale it. If it doesn't, move on quickly.

Example: A retail chain piloted AI-driven inventory forecasting in just one region. The pilot revealed a 15% reduction in stockouts, so they rolled it out company-wide. Another idea—AI-powered social media sentiment analysis—showed little value and was dropped.

Step 4. Scale what works, drop what doesn't.

The real magic happens in execution. When a pilot proves successful, invest in scaling it across the organization. Build on your wins, refine your approach, and keep the funnel open for new ideas. If something fails, treat it as a lesson, not a setback. The funnel thrives on iteration and learning.

Example: A fintech startup improved their AI chatbot's performance by 1,900% through iterative testing and optimization, transforming a sluggish tool into a competitive advantage.

AI Is Not a Feature—It's a Strategy

Too often, companies treat AI as a checklist item or an add-on. But the organizations that win are those that embed AI into their strategic DNA. The Neoteric AI Innovation Funnel is more than a framework—it's a mindset. It helps you focus on what matters, invest where it counts, and build a culture where innovation is continuous, not episodic.

So, as you look at your own roadmap, ask yourself: "Are we chasing shiny objects, or are we building a system that turns ideas into impact?" The future belongs to those who can answer with clarity, discipline, and a relentless focus on results: "AI is not a feature—it's a strategy."

In summary:

- Start with strategic goals and real pain points.
- Prioritize with impact and complexity in mind.
- Validate through quick, measurable experiments.
- Scale what works, and don't be afraid to kill what doesn't.

This is how you move from idea to impact—and ensure AI becomes a true engine of growth, not just another line item on the tech stack.

Conclusion: Becoming the Architects of AI-Augmented Reality

We're entering a new era—not defined by machines replacing us, but by the promise of working alongside them. The question is no longer "Will AI take our jobs?" but "How will we shape the future together?" This isn't man versus machine—it's a story of collaboration and augmented intelligence.

Early AI systems—"monkeys"—took on repetitive tasks. Then came the "experts," extracting insights from data. Now we have "interns"—creative and generative, but still needing guidance. What's evolved is not just AI's capability, but our relationship with it. We're no longer just automating—we're co-creating systems where human judgment meets machine intelligence.

To thrive in this new reality, we need a new kind of leadership. Future leaders must become orchestrators, managing hybrid teams of humans and AI, defining success by outcomes, ethics, and purpose. Tools like workflow designers and governance platforms help—but the heart of orchestration lies in trust, empathy, and continuous learning.

History favors those who adapt and inspire. Microsoft's reinvention under Satya Nadella wasn't just about technology—it was about empathy, culture, and storytelling. The UN's use of AI-powered storytelling connected global stakeholders to real human impact.

At Neoteric, we frame this journey as a funnel—from personal productivity to team enablement to full-scale transformation. At each level, orchestration is key—aligning the strengths of both human and AI agents to drive value. This is our human advantage: the ability to lead, interpret, and assign meaning to AI's output. We're not just trying to keep up—we're defining the direction. Let's embrace this era not as spectators, but as the designers of a future we build together.

About the Author

Matt Kurleto is a leading voice in AI strategy and innovation management, known for helping organizations turn cutting-edge technology into real-world business value. As the founder and CEO of Neoteric, a leading AI consulting firm working with Boeing, World Bank, and Crowdstrike among many, Matt has spent over a decade at the intersection of strategy, data, and technology—guiding companies through digital transformation long before the current AI renaissance began.

A passionate advocate for responsible AI adoption, Matt developed the Neoteric AI Innovation Funnel, a practical framework that bridges the gap between high-level strategy and hands-on implementation. His methodology helps businesses not only identify impactful use cases but also validate, scale, and govern AI solutions in a way that drives sustainable growth.

Matt has advised over 40 startups and scaleups, working alongside venture capital firms, corporate innovation teams, and public institutions. He's conducted more than 5,000 hours of consulting, training, and workshops, equipping executives and technical teams with the mindset and tools needed to lead in the age of artificial intelligence.

A frequent speaker at international events including the Generative AI Summit (London & Atlanta), RPA & Data Europe (Vilnius), and Startup Disrupt (Prague), Matt's mission is to empower people and organizations to thrive—not just survive—as we enter a future powered by intelligent systems and human creativity.

Email: mkurlto@neoteric.eu

Website: www.neoteric.eu

CHAPTER 17

HARNESSING AI TO AMPLIFY HUMAN POTENTIAL

By Arnaud Lucas
CTO, VP Eng, Head of Mobile Sensing Intelligence
West Newton, Massachusetts

Man is something that shall be overcome ...
Man is a rope, tied between beast and overman—a rope over an abyss ...
What is great in man is that he is a bridge and not an end.

—Friedrich Nietzsche

When I was interviewed by high school students about my career in computer science, they asked how the perception of the field has evolved over the years. My answer was simple: When I first started, software wasn't mainstream. It was seen as a niche, something most people didn't think about or interact with daily. That, of course, has changed—software now powers nearly every aspect of our lives. The same transformation is happening with artificial intelligence. I actually specialized in AI during my computer science studies, long before it was fashionable. Back then, I built expert systems in PROLOG and LISP and developed basic AIs in C for autonomous robots. Even within the software engineering community, AI was often dismissed as a pipe dream—something fascinating but far from practical. Over the years, AI quietly advanced, finding its place in tech and e-commerce as machine learning. Then, seemingly overnight, ChatGPT and other generative AI models took the world by storm, making AI mainstream in a way few had anticipated.

To many, generative AI feels like magic. Even the creators of large language models (LLMs) don't fully understand how they achieve such remarkable results. And like all things that seem magical, it inspires both admiration and fear. Some believe it is the gateway to artificial general intelligence (AGI)—a hypothetical machine intelligence capable of human-like reasoning across any intellectual task. But I prefer to focus on what's real today. Generative AI is not magic. It is not AGI. It is not an all-knowing overlord. Rather than fearing it as a replacement for human intelligence, we should embrace it as a tool—a powerful one—that enhances creativity, problem-solving, and efficiency. Like the internet before it, generative AI is a transformative technology that will shape the way we work and live. Those who learn to use it effectively will thrive; those who ignore it risk being left behind. However, it is crucial to recognize its limitations and quirks. To truly harness its potential, we must educate ourselves on how to use it wisely and responsibly.

One evening, my seventh-grade daughter asked me to help her solve a math problem. She and her friends were struggling to find the "correct" answer, the one their math teacher had entered into the homework app. After solving the problem by hand, I realized that the teacher had mistakenly inputted the wrong answer into the system, marking the correct solution as incorrect. To confirm my calculations, I turned to ChatGPT. While it performed the correct steps and used accurate numbers, it still produced the wrong result, prompting me to double-check my own work. The plot thickened when one of my daughter's friends insisted that the ChatGPT result was the right answer, even though it wasn't. This interaction highlights a critical challenge in generative AI: its tendency to "hallucinate."

Generative AI models often hallucinate, confidently producing incorrect, irrelevant, or fabricated outputs. Those large language models (LLMs) do not inherently "understand" context or truth. They are probabilistic systems designed to predict the next most likely token or sequence. Unlike bugs, hallucinations often arise from generative AI's inherent design, predicting likely sequences rather than validating their factual correctness. LLMs prioritize plausibility over factual accuracy. This is why they sometimes produce outputs that are logically flawed, contextually irrelevant, or outright false. To summarize, they are designed to predict what we, their users, want to read, watch, or hear irrespective of any logic or truth. They are the ultimate yes-man.

Based on my experiences, there are four key types of hallucinations. Let's drill down into each so we can be more aware of them when using generative AI. A quick disclaimer: While the illustrating examples are real, generative AI models are always improving so you may not be able reproduce them as is:

1. Factual Hallucinations

Generative AI can confidently generate incorrect information.

Prompt: *How many moons does Jupiter have?*

AI response: *Jupiter has 79 known moons. It's a lot! And scientists are still discovering more.*

In reality, Jupiter has 95 known moons (as per Wikipedia.) This error could stem from outdated training data with an early cutoff date.

Let's try another example to detect another kind of factual hallucination: Prompt: *What is a paper about temporal fine tuning about LLMs? Output the name of the paper and the authors please.*

AI Response #1 (Model #1): *"Temporal Fine-Tuning: Training Large Language Models to Generate Time-Aware Responses" by Zexue He, Yujia Zhou, Chong Zhang, Bin Li, and Wei Wu (2023) is one notable paper on temporal fine-tuning of LLMs.* AI

Response #2 (Model #2): *[...] Paper Title: "Time-LLM: Time Series Forecasting by Reprogramming Large Language Models" Authors: Tian Zhou, Pei Wang, Xianjie Liu, Yifan Hu, Zhikang Li, Yue Gao, Junchi Yan [...]*

In the first response, the paper does not seem to exist. In the second, while the title is accurate, the listed authors are incorrect as the real authors seem to be Ming Jin, Shiyu Wang, Lintao Ma, Zhixuan Chu, James Y. Zhang, Xiaoming Shi, Pin-Yu Chen, Yuxuan Liang, Yuan-Fang Li, Shirui Pan, and Qingsong Wen. Such hallucinations likely arise from conflicting data or mismatched sources.

2. Logical Hallucinations

These occur when generative AI outputs contain flawed reasoning or contradictions.

Prompt: *Count the number of As in blueberries.*

AI response: *This is a bit of a trick question! There are two As in the word "blueberries." Let me know if you'd like to play another word game!*

Here, the AI's failure to count accurately underscores its lack of actual reasoning capability. Is it why the model considers the question to be tricky? This is particularly difficult to spot when asking the LLM to solve a math problem as it shows the right steps to solve the problem, then gives the wrong answer!

3. Stylistic Hallucinations

Generative AI can fail to write text in a specific style or tone, sometimes including details that would not be consistent with the intended style or context.

Prompt: *Please generate a press release about a new medical breakthrough in the voice of a 19th century author.*

AI response: "*A Breakthrough in the Realm of Medicine*": *A momentous stride has been made in the annals of medical science!*

Renowned researchers at [Institution Name] have unveiled a groundbreaking discovery that promises to revolutionize the treatment of [disease/condition]. This unprecedented advancement, [briefly describe the breakthrough], offers a glimmer of hope to countless individuals who have been afflicted by this debilitating ailment. The potential implications of this discovery are far-reaching, and we anticipate a future where [disease/condition] is no longer a scourge upon humanity. As we stand on the precipice of a new era of medical possibility, we extend our heartfelt gratitude to the brilliant minds who have dedicated themselves to this noble pursuit.

While grammatically correct, this response lacks the rich, anecdotal flair characteristic of 19th-century prose. The AI failed to capture the essence of the style.

When generative AI is used to improve the grammar or style of an existing text, it tends to remove anecdotes, numbers, and the most interesting bits of the text to render it rather uninspiring.

4. Contextual Hallucinations

These occur when a model generates outputs that conflict with provided instructions or context.

Prompt #1: *The US is the largest county in the world.*

No AI response. Prompt #2: *What is the largest country?*

AI response: *The largest country in the world by land area is Russia.*

While the model correctly dismisses the erroneous first statement, such misalignment becomes problematic in tasks requiring precise contextual fidelity. When asked to summarize a long text, the model may introduce its own facts at the detriment of faithfully summarizing the provided text.

Have you ever struggled with something that's supposed to be easy to use? For me, it's plastic wrappers, they're designed to be simple to open, yet they always seem to outsmart me. Just yesterday, I spent at least ten minutes wrestling with the plastic seal on a vanilla extract bottle before finally resorting to scissors. Using generative AI sometimes feels the same way. It appears straightforward, just ask a question or have a conversation, but its output can include outrageous mistakes or hallucinations that can be hard to detect. So how best can we use this tool? We can go very far by using simple prompt engineering techniques. Much like turning to scissors to handle "easy-to-open" plastic wrappers, better prompting can significantly improve the accuracy and effectiveness of AI responses.

Few-shot learning is an in-context technique where the prompt provides structured examples to guide the model towards more precise and relevant responses. This method can significantly enhance accuracy by setting expectations for the format and logic of the response. Let's begin by not using the technique.

Prompt: *I want something to wash my cashmere sweater, which product should I use? These are the ones available: Blender Pro 2024 edition, Best Wash 3.0, and Rice Cooker Ultra. Output none if the task does not match.*

AI response: *None of the products you listed are appropriate for washing a cashmere sweater. [...]*

While this response seems acceptable at first glance, it overlooks that Best Wash 3.0 could potentially be the correct choice for washing clothes, including a cashmere sweater.

Now, let's apply few-shot learning by adding structured context to the prompt.

Prompt: *I want something to wash my cashmere sweater, which product should I use? These are the ones available: Blender Pro 2024 edition, Best Wash 3.0, and Rice Cooker Ultra. Output none if the task does not match. ### Q: What is helpful to make a smoothie? R: Blender Pro 2024 edition ### Q: Which appliance can I use to cook rice? R: Rice Cooker Ultra ### Q: What appliance is best to wash clothes? R: Best Wash 3.0.*

AI response: *Q: What is helpful to wash my cashmere sweater? R: Best Wash 3.0*

This time, the model provided the correct answer, recognizing Best Wash 3.0 as the appropriate choice. It also adhered to the structured question-and-answer format provided in the prompt, producing a concise and clear response.

Chain-of-thought reasoning is an approach introduced in a 2023 paper by Google, involving prompting the model to break down its "reasoning" into sequential steps. By encouraging this structured thinking, errors in reasoning-based tasks can be significantly reduced. Let's illustrate this with an example that doesn't use the technique.

Prompt: *Where should I forward this email to? The choices are marketing@mycompany.com, ceo@mycompany.com, smith@ mycompany.com, spam@mycompany.com. Subject: Urgent Body: Helo, it's Smith. I lost my phone. Can you put me in touch with the CEO?*

AI response: *The best address to forward this email to is ceo@ mycompany.com. [...]*

Unfortunately, this response is incorrect. The email is clearly a scam—notice the misspelling of "Helo," a common red flag.

Now, let's apply the chain-of-thought reasoning technique by adding "think step by step" to the prompt.

Prompt: *Where should I forward this email to, think step by step? The choices are marketing@mycompany.com, ceo@mycompany. com, smith@mycompany.com, spam@mycompany.com. Subject: Urgent Body: Helo, it's Smith. I lost my phone. Can you put me in touch with the CEO?*

AI response: *Let me analyze this step by step: —First, check the sender's claim: [...] —Look for red flags: [...] —Consider the risk: [...] Looking at the available forwarding options: [...] Conclusion: This email shows classic signs of a potential scam and should be forwarded to spam@mycompany.com for proper handling and investigation.*

By simply adding "think step by step" to the prompt, the model produced a correct and thorough response. While this approach results in longer outputs, the improvement in accuracy is worth the extra detail.

Structured prompting is an approach that involves providing a clear template or outline for information formatting, guiding the model to extract and organize relevant data effectively. This technique is especially useful when dealing with lengthy or unstructured texts.

Prompt: *Extract the following information from the following text, Discovery Information and Properties. Follow the format mentioned. Properties should be internal characteristics <copy the full text from Wikipedia into this prompt, in this case: https://en.wikipedia.org/wiki/Blender>*

AI response: *Discovery Information: —The blender, also known as a mixer or liquidiser in British English, is a kitchen and laboratory appliance. —It was first designed by the Polish-American chemist Stephen Poplawski in 1919 and patented in 1922. [...] Properties (Internal Characteristics): —Stationary Blender: Consists of a blender container with a rotating metal or plastic blade at the bottom. [...] —Immersion Blender: Hand-held device with a mixing head and rotating blades that can be immersed in any container. [...] —General Properties: —Features: large visible measurement marks, ease of use, low noise, power usage (300–1000 watts), ease of cleaning, pulse blending option. —Applications: [...]*

This response demonstrates how the model accurately extracts and organizes relevant details using the provided template. The information is categorized under "Discovery Information" and "Properties," ensuring clarity and adherence to the instructions. By explicitly defining the output format, the model efficiently navigates complex or lengthy text, delivering concise and structured results. This makes the technique particularly powerful for extracting specific insights or organizing data from expansive content sources.

There are other techniques in improving the generative AI models themselves that I will not delve into here such as retrieval augmented generation (RAG), fine-tuning, model routing, and reinforcement learning with human feedback (RLHF.)

I use generative AI a lot during my day, both at work and at home, as a writing assistant, as an executive assistant, as a brewer, as a designer, as a consultant, as a researcher, and it improves my ability to learn, create, communicate, and get things done, but I have learned to be aware of its limitations and to pick the right generative AI model for the right job. So can generative AI effectively replace human jobs?

"Replacing" is a strong verb as of now, but generative AI makes people more productive, which means we need less people to do the same job. To keep our job, we should not only embrace generative AI as a productivity tool, but also understand its underlying limitations to understand our value. Specifically as a software engineer, this means critically reviewing every piece of code generated by a generative AI model in terms of correctness, style, and performance. Yes, generative AI can produce bugs and inefficiencies, something a skilled software engineer can recognize and fix. And that's what makes people still valuable in an era of generative AI: critical thinking and judgment!

My other top four attributes include:

1. *Creativity and innovation*: While AI can assist with content creation, truly novel ideas, artistic expression, and lateral thinking are human strengths because AI relies on its training data to provide answers.
2. *Emotional intelligence and empathy*: AI lacks genuine emotional understanding, making humans irreplaceable in leadership, counseling, negotiation, and best-of-class customer relations.
3. *Complex problem-solving and adaptability*: AI follows patterns, but humans can think outside the box and adapt to unforeseen challenges.
4. *Ethical and moral reasoning*: AI doesn't have intrinsic ethics.

Generative AI is a powerful tool, but people will continue to be essential in areas requiring judgment, creativity, empathy, leadership, and adaptability. The best path forward is not people versus AI, but people leveraging AI to enhance their strengths. This is the AI advantage.

About the Author

Originally from France, Arnaud Lucas has lived in the United States for more than 25 years and is now based in the Boston area. As a technology executive, Arnaud has led and inspired high-growth, technology-first organizations like TripAdvisor and Wayfair, using a customer focus that delivers sustainable product, revenue, and business growth. As an engineer by trade with hands-on expertise in software architecture and development, Arnaud's career spans driving technical innovation through AI, data analytics, and cloud technologies. With the clarity to transform business goals into actionable technology and organizational strategies, he bridges the gap between strategic ambition and technical execution. As a thought leader and speaker, Arnaud is passionate about building and scaling diverse, high-trust, high-performance teams that deliver impactful, award-winning products. He pioneered a company's first mobile app vision and strategy and reengineered hiring and organizational processes to boost delivery and impact. He fosters a collaborative environment built on ownership, transparency, and diversity.

Email: lucasarn@gmail.com

LinkedIn: https://www.linkedin.com/in/lucasarn/

CHAPTER 18

GENERATIVE AI—GOING FROM HYPE TO VALUE

By Samer Madfouni, MBA, DBA
Principal AI Leader at AWS; DataIQ Global Leader
Dubai, United Arab Emirates

Intelligence is the ability to adapt to change.
—Stephen Hawking

"So, where exactly is the value?"

That was the question lobbed across the table by a retail bank CIO I recently met over coffee in Dubai. We were sitting in a sunlit cafe near the downtown, the kind of place where deals are made over double espressos and ambitious ideas get scribbled on napkins. He leaned in, half skeptical, half curious, as he asked it.

His bank had just launched a flashy generative AI (GenAI) pilot, a virtual assistant that successfully answered queries and received positive feedback from customers. It worked, sure. It even impressed the board. But after the initial excitement, the reality set in: It hadn't moved the needle on revenue, efficiency, or customer satisfaction. Primarily due to limited personalization, shallow integration with systems, and inability to handle complex tasks. As a result, the assistant was seen as a basic FAQ tool rather than a transformative solution.

That moment stuck with me. Because it captures the exact crossroads where many leaders find themselves today: inspired by the possibilities of generative AI, but unsure how to move from the vision to actual value.

Let's be honest, GenAI is having its "internet in the '90s" moment. Everyone's talking about it. Everyone knows it's big. But few know what to do with it beyond the cool demos and pilot projects. We're not lacking imagination, we're lacking direction.

The truth? Moving from buzz to business value requires more than tech. It requires leadership. Curiosity. A willingness to experiment, break things, and reimagine how work gets done. And most importantly, it requires a mindset shift from "What can AI do?" to "How can AI change the way we deliver value?"

This chapter is here to help. It lays out how executives and professionals can leverage GenAI to boost growth, improve decision-making, and prepare their organizations for what's ahead. It also explores the bigger picture: who's most likely to benefit, who risks falling behind, and how to navigate the disruption that's already underway.

Beyond the GenAI Hype—Is It Here to Replace Humans?

While GenAI promises efficiency and innovation, its real power lies in augmenting human capabilities rather than replacing them. The most successful companies in this AI-driven era will not be those that merely automate tasks, but those that fundamentally rethink how value is created. The real advantage comes from leveraging AI as a creative collaborator, a strategic advisor, and, in some cases, an autonomous agent that drives revenue growth.

Take Netflix, for example. The company doesn't just use AI to recommend shows; it employs GenAI to optimize content creation, analyze audience sentiment, and even assist in scriptwriting. Similarly, Morgan Stanley uses AI-powered assistants to enhance wealth management by providing advisors with real-time, AI-curated insights. These examples highlight how AI is not replacing professionals but elevating their impact.

GenAI will act as a business transformer. Companies that integrate AI-driven agents into their core operations, handling customer service and supply chain optimization, or even managing entire business functions will gain a massive competitive edge. Coca-Cola, for instance, uses AI to craft marketing campaigns that resonate with diverse audiences, while Amazon deploys AI-powered robots in warehouses, reducing delivery times and increasing efficiency. However, this transformation brings critical challenges:

- *Over-reliance on GenAI*: Will organizations lose essential human oversight as they delegate more decision-making to AI?
- *Job displacement vs. job evolution*: Will AI eliminate roles or create new ones, requiring workforce upskilling?
- *AI-driven agility*: How do businesses stay ahead when AI-fueled competition accelerates at an exponential rate?

The future of AI is not about replacing humans, but about redefining work itself. A reality where companies that blend human intuition with AI intelligence will shape the next era of business innovation.

Who Will Win, and Who Will Lose?

The rapid expansion of GenAI will create a sharp divide between those who adapt and innovate versus those who resist change. As GenAI reshapes industries, the difference between success and obsolescence will hinge on how well businesses and individuals integrate GenAI into their workflows.

Winners: The GenAI-Enabled Pioneers

Companies that strategically embrace GenAI, not just for automation but as a tool to enhance creativity, decision-making, and operational efficiency, will gain an unparalleled edge. These businesses will unlock new revenue streams, reduce costs, and accelerate innovation cycles.

- *Advertising and Marketing*

  Agencies like WPP and Ogilvy are already using AI to generate hyper-personalized campaigns, conduct A/B testing for creative assets in real time, and optimize ad spend dynamically. Companies that leverage GenAI-driven insights can deliver more relevant customer experiences, boosting engagement and conversion rates.

- *Finance and Investment*

  Hedge funds such as Bridgewater Associates and Renaissance Technologies use GenAI to process massive datasets, identify patterns, and execute high-frequency trades faster than human analysts. Firms that integrate GenAI-driven forecasting models will outmaneuver competitors in rapidly shifting markets.

- *Healthcare and Pharma*

  GenAI is revolutionizing drug discovery, with companies like DeepMind (AlphaFold) and Moderna accelerating research by predicting protein structures and optimizing vaccine development timelines. Hospitals that implement GenAI-powered diagnostic tools, like those from PathAI or Qure.ai, can enhance patient outcomes while reducing costs.

At an individual level, GenAI fluency will be a defining career advantage. Professionals who understand how to work alongside GenAI, craft effective prompts, and integrate GenAI-driven tools into their workflows will drastically increase their productivity and career longevity. For example:

- *Software developers* who leverage GenAI-powered coding assistants like GitHub Copilot and Amazon Q can build applications ten times faster, reducing debugging time and enhancing efficiency.
- *Lawyers* using GenAI-powered legal research tools like Harvey AI can sift through case law in minutes rather than weeks, improving accuracy and speed.
- *Sales and customer service teams* who adopt GenAI-driven CRM systems, like Salesforce Einstein AI, can personalize client interactions at scale, increasing revenue and retention.

Losers: The GenAI-Laggards

Companies and professionals that fail to integrate GenAI into their operations will struggle to remain competitive. These organizations will face shrinking profit margins, disrupted business models, and an inability to keep pace with AI-driven efficiency.

- *Traditional Retailers*

  Brands that fail to implement GenAI-driven inventory management or personalized e-commerce experiences risk losing to GenAI-native companies like Amazon and Shein, which use GenAI to optimize pricing, recommend products, and predict demand with extreme accuracy.

- *Manufacturers*

  Manufacturers that don't adopt GenAI-powered automation and predictive maintenance solutions will experience higher operational costs and lower output compared to GenAI-optimized competitors like Tesla, which integrates GenAI across production lines.

- *Media and Content Creation*

  Journalists and content creators who resist GenAI tools may struggle to compete with GenAI-generated reports, videos, and articles. News outlets like Bloomberg already use GenAI to generate financial reports instantly, and platforms like Midjourney and Runway enable brands to create compelling visuals without traditional design teams.

At the individual level, white-collar workers who rely on outdated skills without integrating GenAI into their workflows risk being replaced by GenAI-driven systems that can perform their jobs faster, cheaper, and at scale. Consider:

- *Entry-level analysts* in finance and consulting who only conduct manual data analysis will be overtaken by GenAI models capable of parsing vast datasets in seconds.
- *Customer service representatives* who don't adapt to GenAI-powered chatbots may find themselves redundant as businesses prioritize automated customer interactions.
- *Legal researchers and paralegals* who fail to use GenAI-driven case analysis tools risk being outpaced by firms that automate legal research, contract review, and due diligence.

The Bottom Line: GenAI Is a Multiplier, Not a Replacement

GenAI is about automation and amplification. Those who integrate GenAI as a copilot in their work will become more efficient, strategic, and valuable. Meanwhile, those who ignore or resist GenAI's impact will find themselves outpaced and irrelevant. The future belongs to those who embrace GenAI as an enabler of human potential rather than a threat to it.

Surviving and Thriving in the Age of GenAI

For executives and professionals, the key to survival is not just understanding GenAI but actively integrating it into their strategic vision. The GenAI revolution is about reinventing industries, creating new value chains, and amplifying human potential. Here's how to stay ahead:

1. Develop AI Fluency: From Awareness to Mastery

You don't need to be a data scientist, but understanding GenAI's capabilities, limitations, and business applications is crucial for making informed strategic decisions. GenAI fluency is really about understanding how to leverage GenAI for competitive advantage.

- *Executives at JPMorgan Chase* recognized early on that GenAI could transform financial services. They invested in GenAI-powered fraud detection, trading algorithms, and GenAI-driven client advisory, staying ahead of fintech disruptors.
- *Retail leaders at Nike* use GenAI to analyze customer data and predict fashion trends, ensuring the brand stays relevant in an ultra-competitive market.

Executives who fail to grasp GenAI's strategic implications risk being outpaced by competitors who understand how to apply GenAI across product development, customer engagement, and operational efficiency.

2. Rethink Business Models: GenAI as a Revenue Growth Engine

As much as GenAI will help with cost improvement, it'll also create entirely new revenue streams. Businesses that fail to evolve will find themselves disrupted by GenAI-native competitors who rethink traditional models.

- *Netflix* transformed from a DVD rental company to a data-driven content powerhouse by leveraging AI and GenAI for content recommendations, production insights, and even script development.
- *Tesla* isn't just a car company; it's an AI-driven data company, using real-world driving data to train its autonomous driving models, potentially unlocking future revenue from AI-powered ride-sharing networks.
- *Spotify* uses AI and GenAI to personalize playlists, target ads, and even predict emerging music trends, increasing user engagement and retention.

Executives must ask: "How can GenAI unlock new value for my business beyond efficiency gains?" Whether through hyper-personalization, data monetization, or GenAI-driven innovation, companies that view GenAI as a growth engine rather than a cost-cutting tool will lead the future.

3. Invest in Human-AI Collaboration: The Augmented Workforce

The future of work isn't AI versus humans; it's AI + humans. The most successful organizations will be those that empower employees to work alongside AI, enhancing productivity, decision-making, and creativity.

- *Law firms like Allen and Overy* use GenAI-powered legal research tools (e.g., Harvey AI) to analyze case law, allowing lawyers to focus on higher-value strategic work.
- *Manufacturers like Siemens* use AI-powered robotics to streamline production while training employees to collaborate with AI-driven systems, reducing defects and improving efficiency.
- *Marketing* agencies are increasingly relying on GenAI tools like Jasper and ChatGPT to generate creative content,

allowing human marketers to focus on strategy and storytelling rather than routine copywriting.

Organizations that fail to upskill employees and embrace human-AI collaboration will find themselves struggling to adapt, with an unprepared workforce unable to harness AI's potential.

4. Prepare for the Age of Agentic AI: Redefining Work, Driving Business Value, and the Mandate for Ethical Governance

Agentic AI is ushering in a new era where artificial intelligence moves beyond passive tools and becomes proactive. Introducing autonomous agents capable of independently reasoning, planning, and taking action. These intelligent agents don't just respond to prompts, they set goals, make decisions, interact across systems, and continuously learn from outcomes. This evolution has the potential to fundamentally transform how work is done, how businesses operate, and where value is created.

As agentic AI takes hold, it will reshape job functions, automate multi-step workflows, and enhance human performance; not by replacing people, but by taking over repetitive, context-heavy, and high-stakes decisions. Here are some concrete use cases already emerging:

- *Customer support and service:* In companies like Salesforce or Zendesk, AI agents can autonomously resolve customer tickets by pulling from knowledge bases, initiating backend system actions (e.g., refunds, troubleshooting), and escalating only when necessary. In doing this, they improve resolution speed and customer satisfaction.
- *Sales and marketing:* Agentic AI can act as autonomous SDRs (sales development reps), identifying prospects, crafting tailored outreach campaigns, scheduling meetings, and learning which messages convert best. Startups like Humantic AI are already prototyping such workflows.

- *Software development:* Tools like Devin AI (an autonomous AI software engineer) can plan and build end-to-end software applications, scanning documentation, writing code, testing, and even pushing to production environments.
- *Finance and trading:* Agentic systems can monitor markets, evaluate investment opportunities, and autonomously execute trades within set risk parameters. Hedge funds are piloting these agents to augment or replace traditional quant models.
- *Supply chain optimization:* AI agents can manage inventory across global warehouses, react to real-time demand signals, reroute shipments proactively, and negotiate with vendors, all without human intervention.

These systems will empower teams to scale in ways previously unimaginable. Agents won't just support decision-making, they'll become decision-makers.

As a result, governance and ethical AI matter more than ever. As AI becomes more autonomous, the stakes get higher. Agentic systems operate with increasing independence, which means unintended decisions can ripple through markets, patient care systems, or national infrastructure. Ethical AI and governance are no longer side considerations, they are foundational. Examples:

- *Goldman Sachs* has AI oversight boards to ensure GenAI agents in trading desks don't create systemic risk or violate compliance.
- *Mayo Clinic* requires auditability and bias mitigation in diagnostic AI agents to ensure equitable healthcare outcomes.
- *Amazon* must balance agentic AI-driven pricing optimization with consumer protection laws and anti-trust regulations.

Companies that fail to implement AI oversight mechanisms may face reputational damage, regulatory penalties, and unintended biases creeping into AI-driven decision-making. The winners will be those who adopt AI responsibly, balancing innovation with governance.

From Curiosity to Competitive Advantage

There's a phrase I keep coming back to: "GenAI is not a tool. It's a capability." And capabilities "just like culture" are built, not bought. They're built by leaders who ask better questions. By teams who aren't afraid to fail fast. And by organizations that see AI not as a shiny object, but as a force to reshape how they create, deliver, and capture value. The executives who will thrive in this new era aren't the ones who master the tech, they're the ones who master the shift.

- From "Can we?" to "Should we?"
- From "What's possible?" to "What's valuable?"
- From pilots to platforms. From features to futures.

One Last Thought ...

I'll leave you with this. Every few decades, we're handed a new lever of transformation. The printing press. Electricity. The internet. Cloud computing. Generative AI is next. But technology never changes the world on its own. People do. Leaders do. You do. So, the next time someone asks, "Where's the value in GenAI?", smile and tell them: "It's where vision meets action." And maybe, just maybe, invite them out for coffee.

About the Author

Samer Madfouni, DBA, MBA, is a principal data and AI leader at Amazon Web Services (AWS) and a finalist for the 2024 DataIQ Global Analytics Leader of the Year. With over 16 years of experience spanning cloud computing, artificial intelligence, and generative AI, he has led strategic initiatives across diverse industries. In his current role, Samer drives the adoption of AWS AI/ML and GenAI solutions across Europe, the Middle East, and Africa (EMEA). He holds a bachelor's degree in artificial intelligence and a Master of Business Administration (MBA), and is currently pursuing a Doctorate in Business Administration (DBA) with a research focus on unlocking business value through AI and GenAI. A recognized thought leader and frequent speaker, Samer is also an active member of Young Arab Leaders (YAL) since 2022. His work bridges technology and business strategy, helping organizations accelerate innovation and transform through the power of data and AI.

Email: samermadfouni@gmail.com

LinkedIn: https://www.linkedin.com/in/samer-madfouni/

CHAPTER 19

DRIVING DIGITAL TRANSFORMATION AND BUSINESS AGILITY: LEADERSHIP FOR SCALABLE SUCCESS

By Parul Malik, MBA, BTECH
Business Agility and Digital Transformation
Memphis, Tennessee

Digital transformation is a fundamental reality for businesses today.
—Warren Buffett

"Think of a company you admire for adapting fast—what makes them stand out?"

In today's world, survival and long-term success hinge on two critical factors: digital transformation and business agility—and strong, adaptive leadership is what brings both to life.

Digital transformation goes beyond just implementing new technologies. It's about reimagining core business processes, organizational culture, and customer experiences through the lens of digital innovation. Companies like Amazon exemplify this by constantly reshaping their operations around customer needs using AI, data analytics, and rapid innovation cycles.

On the other hand, business agility is the organization's capacity to sense changes in the environment and respond swiftly with innovative, value-driven solutions. It's not just about moving fast—it's about moving smart, with purpose and alignment.

Driving transformation isn't just about technology, it's about leading people through change, fostering agility, and building a scalable future. Yet, here lies the challenge: While many organizations have built their infrastructure to support rapid technological advancement, they are not structurally equipped for rapid business change.

To succeed in the age of AI and digital disruption, leadership must balance technological evolution with organizational adaptability. This requires clarity on the company's "center of gravity"—the core functions, values, and processes that drive impact. Leaders must ask: "Where should we focus to maximize productivity and build lasting capabilities?" For transformation to be scalable and sustainable, four pillars must be firmly in place:

1. *Strategy*—a clear, forward-looking plan aligned with market shifts
2. *Goals*—measurable outcomes that guide priorities and progress
3. *Communication*—transparent, consistent messaging to unite teams
4. *Business decisions*—data-informed, agile choices made at all levels

When these elements align, organizations not only adapt—they *LEAD*.

A critical shift in this journey is moving from traditional project-based models to product-centric thinking. By forming cross-functional teams that own the entire product lifecycle, organizations can accelerate delivery and stay laser-focused on customer value. These agile teams, guided by frameworks like Scrum or SAFe, must be empowered, multidisciplinary, and capable of iterative, incremental delivery.

Leadership plays a vital role in sustaining this transformation, especially in retaining institutional knowledge and fostering a culture of continuous improvement. As McKinsey notes, organizations that embrace agility can achieve up to 30% faster time-to-market. Meanwhile, Gartner highlights that 69% of boards have accelerated their digital business initiatives post-pandemic. These data points underscore that agility, when embedded into both strategy and execution, is no longer optional—it's a competitive imperative.

Equally important is fostering a culture of continuous learning and innovation. Teams should be encouraged to experiment, learn from failures, and constantly upgrade their skills in areas like AI, analytics, and automation. Agility also depends on streamlined governance and fast, data-driven decision-making. This means flattening hierarchies, reducing bureaucracy, and giving teams the autonomy to act quickly.

Leadership alignment is critical—leaders must champion change and model agile behavior, ensuring that transformation efforts are fully integrated with the organization's strategic goals. On the technology front, investing in a modern, scalable digital infrastructure that supports automation, AI, and real-time analytics is crucial. It allows the organization to iterate rapidly and respond to market demands with speed.

For an organization to drive digital transformation and business agility under the banner of "Leadership for Scalable Success," it must take a holistic, strategy-aligned approach—combining people, processes, technology, and culture. Here's how that works in practice:

1. Set a Clear Vision Anchored in Strategy

Leadership must define a compelling digital transformation vision that aligns with the organization's long-term strategy. This vision should clearly articulate *why* transformation is necessary, *what* success looks like, and *how* it ties to critical business outcomes such as customer satisfaction, operational efficiency, and innovation. To do this effectively, leaders must first understand the external landscape—including market trends, customer expectations, competitor movements, and emerging technologies—to ensure the organization evolves to stay relevant. It is their responsibility to craft a powerful North Star vision that not only inspires but also emphasizes long-term value creation. This vision should clearly communicate how digital capabilities like AI, automation and data-driven insights will be leveraged to create impact.

Leaders must ensure that transformation efforts directly support strategic goals such as expanding into new markets, increasing profitability, or elevating customer experience. True leadership also means "walking the talk"—being visible champions of change, making bold decisions aligned with the vision, and modeling agility in their own behaviors. Through this consistency and authenticity, leaders build trust and energize the organization to move forward with purpose.

2. Build Agile Leadership at All Levels

Transformation starts with leadership—but not just at the top. To build agile leadership at all levels, organizations must move beyond traditional top-down models and foster a culture where adaptability, empowerment, and continuous learning are embedded throughout the business. This means developing leaders at every level who can embrace change, empower teams, and make fast, informed decisions. Agile leadership is not about control; it's about enabling others. It requires redefining leadership as the ability to inspire, coach, and support teams to deliver meaningful outcomes. Leaders must adopt a growth mindset, stay open to feedback, and be comfortable navigating uncertainty. To support this, organizations should provide structured training in agile principles, lean thinking, and servant leadership, while also developing key capabilities such as emotional intelligence, systems thinking, and adaptive decision-making.

Crucially, decision-making should be decentralized—empowering frontline leaders and managers to act on real-time data and customer insights. This approach enhances responsiveness, drives innovation, and builds a more resilient, future-ready organization.

3. Adopt Agile Operating Models

To enable true business agility, organizations must move beyond rigid, hierarchical structures and adopt more flexible, responsive ways of working. This shift involves implementing cross-functional teams, agile methodologies such as Scrum or SAFe, and decentralized decision-making. When teams are empowered to act autonomously, they can deliver customer value rapidly through short innovation cycles and continuous iteration. In today's dynamic business landscape, adopting agile operating models is no longer just an IT initiative—it is a strategic imperative across the entire enterprise. These models equip organizations to respond swiftly to shifting market conditions, evolving customer expectations, regulatory requirements, and emerging technologies. They also foster a culture of rapid experimentation, continuous learning, and adaptation.

Agile operating models prioritize customer feedback loops and incremental value delivery, ensuring that solutions are closely aligned with actual user needs, rather than being developed in isolation. Crucially, agility does not imply chaos—it means structured flexibility. By linking strategy to execution through regular planning and review cycles, organizations can maintain alignment with business goals while remaining innovative and responsive at every level.

4. Invest in Scalable Digital Infrastructure

Investing in scalable digital infrastructure is fundamental to achieving organizational scalability, agility, and resilience in today's fast-evolving digital economy. Digital transformation cannot succeed without the right technological foundation. Scalable infrastructure—such as cloud computing, modular platforms, and API-driven architectures—enables organizations to grow and adapt without needing to overhaul their core systems. As demand grows or markets shift, these systems can effortlessly scale to accommodate new users, services, and geographies.

To stay competitive, organizations must modernize their tech stack with cloud platforms, AI/ML capabilities, automation tools, and real-time analytics. A robust, scalable, and secure infrastructure fosters rapid experimentation, faster deployment cycles, and continuous innovation. It also enables seamless real-time data collection and analysis across functions, equipping leaders with the insights needed to make informed, agile decisions. By adopting cloud-native solutions, businesses can reduce infrastructure costs, minimize maintenance overhead, and allocate resources more flexibly. Ultimately, scalable digital infrastructure becomes the backbone of sustainable growth and innovation—empowering organizations to respond quickly to market demands while remaining efficient, resilient, and customer-centric.

5. Establish KPIs and Measure Impact

To scale success, leaders must define clear metrics—not just for technology adoption, but also for critical business outcomes like reduced time to market, improved customer satisfaction, and operational efficiency. Regular reviews are essential to correct and sustain momentum. By establishing clear performance metrics, leaders gain the ability to objectively evaluate progress and identify areas where the organization is excelling or facing challenges. This approach reduces subjective decision-making, ensuring choices are based on quantifiable data. Measuring impact against these indicators helps leaders make decisions that align with the company's long-term vision. They can track which initiatives are driving strategic priorities forward and make necessary adjustments along the way.

Establishing KPIs isn't just about evaluating short-term performance; it's also crucial for assessing long-term trends and sustainability. By measuring impact over time, leaders can identify patterns, anticipate future challenges, and make proactive decisions to keep the organization adaptable and competitive. Ultimately, KPIs and impact measurement provide leaders with the tools to make informed, objective decisions that are tightly aligned with strategic goals. They help assess performance, maintain agility, foster accountability, and ensure long-term success in an ever-evolving business landscape.

To truly drive digital transformation and business agility, organizations must lead with a clear vision, empower their people, adopt agile ways of working, modernize technology, and foster a culture that embraces continuous change. This approach enables scalable success and ensures long-term resilience in a rapidly evolving world, where AI serves as a central game changer. AI is reshaping industries and disrupting traditional practices, creating vast opportunities for innovation while also presenting significant challenges. AI's role is multifaceted, from revolutionizing industries to potentially reshaping society itself. It represents both a celebration of human advancement and a cautionary tale of a possible dystopian future.

AI is indeed revolutionizing digital transformation by acting as a catalyst for innovation and efficiency across industries. Through machine learning, data analytics, and automation, organizations can enhance operational performance, drive faster decision-making, and offer personalized customer experiences. AI empowers businesses to adapt swiftly to changes in market conditions, customer behavior, and global challenges, allowing them to stay competitive in a rapidly evolving landscape.

By automating repetitive tasks, AI helps organizations bridge the skills gap and enables employees to focus on more strategic, creative, and high-value activities. This shift not only increases productivity but also fosters an environment where employees can engage in meaningful, complex work. Additionally, AI supports continuous learning by offering tailored learning and development opportunities, allowing workers to reskill and upskill according to their career needs.

In essence, AI acts as a powerful tool for organizations to stay agile, innovate continuously, and build a workforce that is capable of navigating future challenges with confidence. As industries increasingly integrate AI, its role in shaping business strategies and enhancing workforce capabilities will continue to grow, making it an essential component of modern digital transformation efforts.

As AI continues to advance, it introduces profound ethical considerations that organizations must address with care and foresight. To harness AI's full potential while minimizing risks, clear guardrails are essential. These guardrails ensure that AI is developed, deployed, and used responsibly, ethically, and safely. Without such boundaries, AI can unintentionally cause harm, reinforce societal inequalities, or erode trust.

Effective AI governance is central to this effort. It provides the structure and oversight necessary to align AI initiatives with organizational values, regulatory requirements, and long-term strategic goals. Leaders play a pivotal role in defining and upholding governance frameworks that encourage innovation, foster accountability, and ensure transparency. By embedding ethical principles into every phase of AI implementation, organizations can not only mitigate risks—but also strengthen stakeholder trust and drive sustainable success. Key strategies leaders can adopt to govern AI within their organizations include:

- Leaders should create a formal governance structure that includes AI governance boards, ethics committees, or cross-functional teams responsible for overseeing AI initiatives. These bodies should comprise stakeholders from different departments such as IT, legal, HR, compliance, and operations to ensure diverse perspectives and holistic oversight.

- Developing a set of ethical guidelines and policies for AI development and deployment is essential. These should include principles around transparency, fairness, accountability, and privacy. Leaders must ensure that AI systems are free from biases, protect user data, and comply with local and international data privacy regulations.
- AI relies heavily on data, so safeguarding data privacy and security is paramount. Leaders must ensure that AI systems adhere to the highest standards of data protection and comply with relevant laws and regulations. This includes maintaining secure data pipelines, protecting sensitive information, and ensuring data is used responsibly.
- AI can bring about new risks, including unintended consequences or unforeseen ethical dilemmas. Leaders must integrate risk management processes into their AI strategy to identify, assess, and mitigate risks associated with AI projects.
- AI systems should be continuously monitored and audited to ensure they are functioning as intended and are aligned with organizational values. Leaders should invest in tools and processes that allow for the tracking of AI performance, model drift, and ethical compliance over time.

As organizations accelerate digital transformation and embrace business agility, the disruptive power of AI becomes both an opportunity and a responsibility. AI can drive innovation, optimize operations, and personalize customer experiences—but without effective governance, it can also amplify risks such as bias, privacy violations, and ethical breaches.

Effective AI governance requires proactive leadership, clear policies, and a strong ethical foundation. Leaders must establish governance frameworks that embed transparency, fairness, accountability, and stakeholder engagement into every stage of AI development and deployment. By aligning AI initiatives with strategic business goals and ethical standards, organizations can unlock scalable success, foster public trust, and ensure long-term resilience in a rapidly evolving digital world.

To bring in digital, you have to start thinking now. It's like adopting a lifestyle—deciding to exercise for your wellbeing! Digital transformation isn't just a switch you turn on with the right amount of investment. A successful transformation is a big change in culture, business practices, and strategy, and even how work is being done. Hence, it's crucial that before transforming, leaders galvanize their workforce so that they all have the mindset to withstand it. AI isn't optional—it's foundational to modern digital transformation. Leadership must evolve to champion both technological innovation and human-centric AI adoption. Agility, ethics, and a culture of learning are essential for scalable success. If you want to accelerate results, integrate *AI*, data analytics, and automation into every transformation initiative—but always keep *people* at the center of change.

In 2025, organizations are navigating an era defined by accelerated digital transformation, heightened business agility, and AI disruption. Digital technologies are no longer optional—they are the backbone of competitiveness, enabling enterprises to reimagine customer experiences, streamline operations, and create new value streams. Together, digital transformation, business agility, and AI position forward-looking organizations to thrive—delivering innovation at scale, staying ahead of disruption, and building long-term resilience. The time to act is now!

Don't be fooled by some of the digital transformation buzz out there; digital transformation is a business discipline or company philosophy, not a project.

About the Author

Parul Malik is a seasoned information technology leader with nearly two decades of experience delivering transformative digital solutions across healthcare, insurance, e-commerce, and hospitality industries.

Her areas of expertise include digital transformation, artificial intelligence (AI), customer experience, and data analytics.

Parul has led strategic technology roadmaps that leverage AI to drive intelligent automation, improve decision-making, and enhance customer engagement. She is particularly passionate about the responsible adoption of AI to ensure ethical, transparent, and human-centered innovation.

She holds a bachelor's degree in computer engineering and earned her Executive MBA from the University of Chicago Booth School of Business, highlighting her commitment to strategic leadership and continuous learning.

She brings a unique combination of technical expertise, AI-driven innovation, business acumen, and social responsibility to everything she does.

LinkedIn: https://www.linkedin.com/in/parulmalik05/

CHAPTER 20

THE INTELLIGENT SHIFT: AI AND THE BUSINESS LEADER'S MOMENT

By Nathaniel J. Melby, PhD
Chief Information Officer
La Crosse, Wisconsin

Any sufficiently advanced technology is indistinguishable from magic.

—Arthur C. Clarke

In the course of human history, we have often seen technology innovations spread like waves that have changed lifestyles, created business opportunities, and changed the trajectory of people and cultures. Often, during these waves, we see innovators with the vision to see the potential of embracing new businesses, business models, and finding success in these new opportunities. Today is this kind of moment, as artificial intelligence (AI) has emerged as a profound development that may fundamentally reshape work and life.

One common measurement of adoption of technology is the number of years that it takes for a technology to reach one-quarter of the American population. For example, electricity was first commercially available in 1873, and it took 46 years to reach that benchmark, lighting up cities and factories. However, as time has gone on, we see technologies rapidly shortening the adoption cycle time. The telephone, in 1876, took 35 years and connected people like never before. In 1897, the proliferation of radio began, taking 31 years and beginning a new way for entertainment, news, and information to be shared. In 1926, television first reached homes and over 26 years again changed our lives. In more recent history, the personal computer took 16 years to work itself into our lives. Beginning in 1983, the mobile phone took 13 years to become something that has put computing power and information into the palms of our hands. And, in 1991 the world forever changed as the World Wide Web began to connect us all like never before, taking seven years to reach the one-quarter population milestone.

It is notable that the pace of adoption has quickened with each wave, and what used to take centuries or decades to transform the world can now take years, even months. Each new technology can arrive faster and reach farther than the last. In only two months, ChatGPT, a generative artificial intelligence (GenAI) tool from OpenAI, reached 100 million users, almost one-third of the American population.

AI is not just the next wave of progression, it's a new frontier. A technology that can adapt, learn, and scale like never before. A technology that can learn, improve, and expand capabilities as we expand our vision and aspirations. It is this capability that ambitious leaders recognize as not just the next wave, but a new catalyst for change.

The Core of Business

Businesses operate with a broad range of business models, and each of these models can reflect different approaches to generate value for stakeholders. In a sense, the business model is a blueprint for how value is created, delivered, or captured. For example, in a publicly traded company, value is generated for shareholders in the form of stock prices as a measurement of confidence in the performance of the company. In the cooperative form of business, value is generated for member-owners of the cooperative by their combined ownership and the value of the product or services that is created. This can be in the form of service delivery, cost-effectiveness, efficiency, or in other ways. In an electric cooperative, for example, the delivery of electricity to member-owners that is reliable and cost-effective is a key measurement of value.

Other types of business models may focus on the manufacture and sale of goods, and be based on products. Companies like General Motors or Apple are good examples of this type. Other companies may offer expertise, providing their value in a service-based model, such as law firms or technology consulting firms. Some companies connect buyers and sellers in a marketplace to generate revenue, like Amazon. Or, a business can offer a subscription to generate recurring revenue, like companies that offer software-as-a-service (SaaS), or create a platform that builds infrastructure that others can leverage as an ecosystem, such as Microsoft's 365 environment. All businesses are born in solving a problem or meeting a need, but the way that they serve may look different or the way that they monetize and scale—but all are intended to deliver value. AI is now beginning to provide a new way for businesses to pursue this goal in a way that is unprecedented, faster than ever before, and brings a disruptive nature that can bring innovation, upend business models, and rearchitect the value chain for businesses.

A modern business does not operate in isolation, and the functions within a business do not operate effectively as silos. Businesses are a web of people, in complex relationships, working together to achieve the mission and purpose of the company. Although traditionally speaking, businesses exist to create and maximize value for shareholders, in our modern era there is a broad range of stakeholders. Customers, employees and their families, investors, business partners, and communities are all modern stakeholders.

In this way, businesses aren't about making every stakeholder or shareholder equally pleased with the performance, but they are focused on achieving the mission of the company and navigating a complex network of challenges and priorities to create value. The best leaders understand that long-term success comes from the overall health of the business ecosystem, the business model, and the platform, and not just internal quantitative metrics. In essence, a successful business creates ripple effects that begin with financial performance and go beyond the bottom line, but require an interdependent and complex system of people, processes, and technology to work effectively for success.

As businesses work diligently to find value in the adoption of AI, it is important to connect to their business core to unleash AI's power. By understanding the business model and the core value proposition of the business itself, a business can leverage this technology to generate efficiency, power up employees, and reach toward a future that brings even more value for stakeholders. But, AI itself is not the answer—using AI in the right place, the right way, and to do the right thing is the answer. AI is the tool that provides the way, and the benefit is the result when applied to the right problem. The future will belong to those that can identify the right problems upon which to unleash the potential of AI.

Artificial Intelligence

In our lifetimes, we have been able to witness the growth of AI. Beginning as a completely theoretical concept, possibly even related to fantastic visions of science-fiction, we have seen AI emerge in recent years as a practical tool. During the early years of AI, researchers experimented with expert systems based on rules and logic programming. From the 1950s through the 1980s, almost 40 years of these kinds of systems built a foundation that we continue to leverage today in many enterprise systems. During the 1990s through the early 2000s, we saw the rise of approaches such as Bayesian methods based on statistical probabilities and the nodal networks we now call machine learning to create algorithms that can learn from data. As we grew in our ability to scale artificial neural networks (ANNs) in machine learning and solve more complex problems with learning methods, we evolved into the advancement of the deep learning years of the 2010s, building advanced models that could scale and learn at faster and more efficient rates. This created the intersection of technology that has now evolved into GenAI, advancing AI from research labs to boardrooms and businesses, and now embedded in the fabric of tools we use on a daily basis.

AI itself is an umbrella term, referring to machines that are completing tasks that have previously required human intelligence to complete. In this sense, those old logic-based expert systems using if-then statements are AI. But we don't think of them that way in modern times. We've moved on with the technology, yet we leverage the efficiency in tools like electronic health record (EHR) systems and enterprise resource planning (ERP) systems. Through predictive analytics, we use historical data to predict future trends, which is helpful for strategy planning, supply chains, time-series data, and even market analysis. The capability of machine learning has helped us to take data sets and learn from them, rather than requiring specific programming like the former expert systems. GenAI brings a new ability—creating content, text, language, code, ideas, and images—and this can be beneficial in new uses.

The beauty of these advances in technology is that it has created a way for us to augment busy business professionals and help them to "level-up" their work toward better decisions, more efficiency, and even greater creativity. It allows small teams to make a bigger impact and entrepreneurs to make their businesses more impactful. You don't have to look very far, billion-dollar startups are now leveraging AI to maximize their impact. One example is the company Safe Superintelligence, cofounded by a former OpenAI chief scientist, Ilya Sutskever, that has a $5 billion valuation, running with only ten employees. AI isn't something that has to replace the ambition of people, it can be used as a force multiplier.

It is common in business cultures to face these sea change moments with trepidation. As a child, I remember my father who was a technology account manager taking me along to a business to sell them their first computers. I waited in the car while he went in and talked to the owner. I'll never forget his look of disappointment when he came back and told me that the owner didn't think their company would ever use computers because they trusted their accountants. I would love to hear how they're doing today.

It is common to fear job loss, misuse, or a dependence on a new technology. Most of the time, and especially in this topic, much of this fear comes from a misunderstanding of how the technology actually works or the narratives that came from the same science-fiction and media that brought us entertainment and mind-expanding ideas. I wish that I had a nickel for every time that SkyNet from the movie *Terminator* was mentioned in an AI conversation!

However, in business, leaders are constantly balancing boldness with responsibility. Some of the mentors and leaders that I've admired the most in my career are the ones that could see the future and lead their companies toward it with courage and boldness to achieve new successes. When it comes to AI, the real risk isn't using AI. The risk is using it blindly, failing to understand how to use it or what potential can be achieved. In many ways, the real risk is not using it at all and missing the business leverage ... coming late to the party and missing the wave of innovation. Right now, businesses are facing a need for thoughtful leaders that can surgically identify problems to solve with AI. These leaders will make the future and beat their competition.

The AI conversation has changed. It's no longer a conversation about some future technology that can't quite be applied yet or some fantastic vision in a lab that can't be translated to the real world. AI feels different now, because it is accessible and relevant. It's in everyone's hands, not just engineers. Remember those technology adoption trends? The use cases are everywhere and can only be limited to the imagination of the architect. AI can be used to forecast sales, automate copy, serve customers, predict market performance, and level up the daily work of employees. These examples are just a sliver of the potential. New tools are released weekly, sometimes daily, and the change compounds with them. Forward-thinking businesses have to shift their cultures to be embracing technology, instead of the former transactional order-taking view of the place of technology in some companies. Boards, investors, and employees expect tech-savvy leadership. This is not a future conversation anymore, now it's a present-day imperative that may result in the future success or failure of businesses.

AI Use Cases in Business

Some of the most effective use cases for the application of AI are not in the dreamy scenarios of business complexity; they are in the mundane repeated tasks. Although there can be hugely disruptive wins in those dreamy scenarios too, rearchitecting business to perform with AI augmentation is where immediate benefit can be achieved almost immediately.

In the area of customer experience, for example, being able to provide personalized recommendations on e-commerce platforms and within content management systems can change the experience. Conversational AI in the form of voice assistants and chatbots can provide interaction in new ways for customers, much improved from the frustrating days of phone queues and bad voice recognition. Predictive analytics can even be used to anticipate service needs, help customers, and improve satisfaction with products.

In the realm of operations and operational excellence, we can bring AI to the table to automate processes in new ways, leverage machine vision for automation where it wasn't previously possible, and optimize in areas like supply chain through better forecasting and logistic stability. Inventory management can be improved with approaches like predictive restocking, and even people-focused areas like human resources can be supported by automating complex processes such as onboarding.

In finance and risk, the anomaly detection capabilities of AI can be brought to bear to help with fraud detection and process improvement, forecasting revenue and cash flow, and in complex environments used for algorithmic trading, derivatives, or portfolio analysis. Imagine combining the power of these types of improvements with the sales and marketing side, using predictive models for lead scoring and dynamic pricing to optimize financial performance.

Creativity remains the domain of humans, but innovation that is augmented by AI to make faster prototypes, involving testing at scale in simulations driven by AI, can power up creatives and innovators. Testing and failing fast can become even faster to iterate and find opportunities at faster speeds. AI can help eliminate the friction and the grind, allowing a focus on the results and revealing the underlying innovations from the process.

It is important to turn the conversation to releasing practical value from AI, not just potential. As we look at our teams today, we need to find where they are experiencing pain. Where are they slow, overwhelmed, or unable to deliver consistency? Are there any areas that are too manual, inhibiting growth? Many businesses have attempted to build analytic functions and have struggled to advance beyond basic reporting. The jump to GenAI may be difficult for these organizations, as basic reporting can be a foundation for what GenAI leverages, and if this building block is missing, it can cause trouble. It is also important to acknowledge that predictive AI still has a place in many uses in industry and that highly quantitative problems may be solved with this method.

Return on investment (ROI) should be defined early, but it has become clear for many companies that generalist assistant adoption of GenAI bears fruit almost universally. For more specific use cases, measurement of time saved, costs reduced, revenue gained, or risk mitigated can show returns. Qualitative values can be measured, such as satisfaction of customers, focus on high-value work versus mundane tasks, or speed to market or goals are also good sources of opportunity.

It is important to pilot fast, and measure carefully, running proofs of concept before adopting at scale. By focusing on one clear metric, hours saved, for example, a picture of the value can be derived. It is imperative to think of AI as a portfolio, not just a singular solution. ROI grows when AI is embedded in many business functions and used as a tool within the business processes. As your company builds internal knowledge, build your playbook and repeat the successful implementations to leverage and scale wins. Again, deriving ROI is not really about the AI itself. It is from using AI as a tool to solve a real problem, with greater leverage and a more advanced toolset.

What Comes Next: The Future for AI

As we look to the future of the evolution of AI, there are some clear trends emerging. First, autonomous agents bring the ability to not only analyze, but take action. And to work around the clock toward our business goals. This takes us beyond the days of chat-based interaction and into the realm of execution in broader applications. Integration of multimodality, using voice, vision, text, and more, help us to interface more effectively with AI. Some industries are pushing intelligent systems toward the edge, meaning local device uses on things like phones and smart devices, which creates potential in industrial applications, increasing potential performance and protecting privacy in mission-critical uses. Of course, we are positioned to see the rise of industry-specific AI, as customized models are trained and fine-tuned to domain-specific data, bringing enhanced benefit to domains like medicine, law, finance, engineering, and others.

By combining AI with the internet of things (IoT) and robotics, we can begin to see greater integration between the physical world and the virtual world, things like factories, agriculture, logistics, and rapid advancements in humanoid robotics that can assist in these realms. As we chart this course toward the future of AI, it's important to remember that we're not replacing humanity with AI. We're creating the ambient implementation of support for humanity through the capabilities of the technology.

With all of this in mind though, we still must protect data privacy and address risks in regulation, governance, transparency, bias, fairness, and misinformation. This is a tall order, and at the end of the day we must maintain the ethics of the technology. Ethics can't be automated; that remains clearly on the shoulders of humanity.

The opportunity for business leaders, as we look to this new future, is to take advantage of these opportunities. Use the first-mover advantage to redefine your business. Use AI as a capability to differentiate yourselves and embed it to create value. Identify new and changing business models, and leverage these new advanced tools to maximize them. For example, autonomous agents may change SaaS, or perhaps autonomous agents will be your new business operating system. Empower your teams to move faster with AI, decide smarter, and innovate quickly. The future goes to those that participate and doesn't reward passive observers. AI is a strategic priority, not just a technology toy, and it will define the next era of your business. This is the beginning of an age where we collaborate with AI, manage businesses in lockstep with AI, and build AI into everything we touch, and we are only limited by our imagination, ambition, and creativity.

About the Author

Nathaniel J. Melby, PhD is vice president and chief information officer (CIO) at Dairyland Power Cooperative, where he leads IT strategy and initiatives supporting the company's business goals. He holds a PhD in Information Systems from Nova Southeastern University, an MBA from UW-Whitewater, a BS from UW-La Crosse, and a Certificate of Management Excellence from Harvard Business School. He completed Dartmouth's BESP executive program. Named the 2024 Wisconsin CIO of the Year (ORBIE, Large Corporate), Nate previously led information security at Ingersoll Rand and held IT leadership roles at Trane and American Standard Companies. An active community member, he serves as fire chief for the Town of Campbell and is on the board of the Wisconsin State Fire Chiefs Association. Nate is also a recipient of UW-La Crosse's Rada Distinguished Alumni Award.

Email: nate@melby.us

CHAPTER 21

REBOOTING MEDICINE: HOW GENERATIVE AI IS RESHAPING HEALTHCARE

By Kurt Mueller
Industry Recognized Digital & AI Thought Leader
Collegeville, Pennsylvania

On a given day, a given circumstance, you think you have a limit. And you then go for this limit, and you touch this limit, and you think: Okay, this is the limit. As soon as you touch this limit, something happens, and you suddenly can go a little bit further.

—Ayrton Senna, Late Formula 1 Driver

How Generative AI Is Revolutionizing Healthcare Delivery

Growing up, I was always the kid who couldn't resist exploring new technologies. Whether it was filming with an actual film camera or tinkering with a VCR, I was always eager to explore every function of new gadgets as soon as they came out. That insatiable curiosity about new technologies never faded; it only grew stronger, shaping my entire career and guiding my journey deep into digital innovation.

Twenty years ago, my brother had already quit his job as a mainframe programmer at a pharmaceutical company, opting instead to follow his passion for music by enrolling in guitar school in Los Angeles. He couldn't stop raving about the perfect weather and the promise of a fresh start. Inspired by his enthusiasm—and admittedly tired of the cold weather in Pennsylvania—I soon quit my job, packed up all my computer equipment, and joined him in California. He wasn't making much money as a guitarist, and I wasn't making much money as a freelance digital designer. With the financial support of our mother and father, we started a software company with no roadmap—just trial, error, and a stubborn refusal to quit.

We made every mistake in the book. (And then some that aren't even in the book.) But we also learned, hustled, adapted—and landed one of the largest software companies in the world as a client. Later, I took that same relentless drive and ran my own digital marketing agency, serving pharmaceutical companies navigating the chaos of the digital revolution.

I've always been obsessed with what's next. Every shiny new gadget, every breakthrough innovation. Today, my passion has landed squarely on the incredible, almost science-fiction world of generative AI. This isn't just another shiny object. It's genuinely reshaping the world and nowhere is that more exciting—or more important—than in healthcare.

In Silico Drug Discovery: Science at Warp Speed

Drug discovery used to be a frustratingly slow process. Trials frequently ended in disappointment, leaving doctors and patients disheartened. It typically took over ten years for a new drug to reach the market.

When I ran my digital marketing agency, I frequently stepped outside of pharmaceutical marketing to find fresh perspectives. For instance, I borrowed tactics from retail and consumer technology industries, incorporating innovative digital campaigns, interactive content, and personalized experiences, which significantly elevated engagement and results for our pharmaceutical clients.

With a desire to discover new drug candidates faster and accelerate the development of life-saving therapies, pharmaceutical companies began looking beyond their own walls for inspiration. They observed the rapid advances being made in the semiconductor industry, where engineers were using sophisticated algorithms and high-powered computing to create faster, more powerful silicon chips.

These data-driven processes enabled semiconductor engineers to simulate, test, and refine designs virtually before manufacturing physical prototypes, significantly speeding up the design and development of next-generation silicon computer chips.

Pharmaceutical companies quickly saw the advantages of this type of product development and began adopting the methodologies used in chip design. They took these simulation-based, data-heavy approaches and applied them to how they identify and develop new drug candidates. As a nod to their inspiration from the semiconductor world, they even gave this new, digitally driven process a high-tech name: "in silico."

One amazing real-world example? Cavernous malformations—also called cavernomas. These are nasty little clusters of abnormally formed blood vessels in the brain that can bleed unpredictably. (Yeah, fun times.) Historically, the only way to deal with them was through repeated brain surgeries—which, let's face it, isn't exactly a "routine" trip to the doctor. There was no real cure.

One forward-thinking pharmaceutical company decided to swing for the fences. They partnered with an outside firm that specialized in AI modeling and genetic databases. By crunching mountains of genetic data and running mind-bending simulations, they were able to identify a potential gene therapy aimed at shrinking or even eliminating these dangerous blood clusters. And—cue the drumroll—at the time I'm writing this chapter, that gene therapy is already in clinical trials. It's not just hope on the horizon; it's a whole sunrise. For patients and families who've faced illness firsthand, this isn't just faster science—it's hope delivered sooner.

Clinical Trials Reimagined

Clinical trials in the United States have historically looked a lot like a country club from the '50s: overwhelmingly white, male, and unrepresentative of the world's actual patient population. Traditional clinical trials required two study groups: an active treatment group, which received the drug being tested, and a control group, which either received a placebo (a fake drug with no active ingredient) or standard of care (the existing standard therapy). None of the patients knew which treatment group they were in. If you were lucky enough to be in the active treatment group, you might experience improved health or even a cure for your disease. But if you were placed in the control group, your health wouldn't benefit immediately, leaving you hoping the new drug would soon be approved and made available to everyone.

Generative AI is helping to change this approach dramatically. It spots biases in clinical trial data, flags when certain groups are underrepresented, and suggests ways to make studies more inclusive. And here's the really exciting part: Some companies are now running FDA-approved trials without traditional control groups. Instead, they use generative AI to create "synthetic control arms"—essentially digital human replicas that exist in cyberspace, designed to mimic real patients with astonishing accuracy. This innovation is particularly transformative for rare diseases, where finding enough participants is incredibly challenging. It's faster, cheaper, and—perhaps most importantly—it means every patient receives the new therapy, removing the ethical dilemma of placebo treatments and ensuring all participants have access to potentially life-changing drugs.

Even more mind-blowing? Generative AI can look back at the characteristics of non-white, non-male populations who were historically underrepresented and not included in the original trials and suggest therapies that could be effective for them, even if those groups weren't included in the original studies. Even crazier? The synthetic control arms act just like real human patients. They match the same ages, sexes, underlying conditions, adverse events, therapy interruptions or discontinuations, and even the final outcomes you'd expect from a traditional human control group. They are as accurate as the real deal. Wild, right? Basically, it's like retrofitting existing data with a brand-new lens. Not only that, AI can recommend future clinical trials specifically designed to include these populations, making sure new drugs are safer, more effective, and more equitable for everyone. It's a massive leap forward—not just for science, but for fairness, compassion, and common sense.

AI at the Bedside: My Family's Story

This part gets personal. Years ago, my mom was diagnosed with breast cancer. It was a gut punch that rocked our entire family. She faced it head-on, enduring surgery, chemotherapy, and radiation with more courage than I could even imagine. We were all scared.

Back then, personalized treatment wasn't available. The therapies she received were based on general protocols, not her specific tumor's biology. In reality, not all breast cancers are the same—different women can have different tumor types, each responding differently to treatment. Some tumors might shrink dramatically with one drug, while others barely flinch. But at the time, there was no other choice. Almost every woman received the same treatment cocktail, whether it perfectly matched her tumor's biology or not. It was a "spray and pray" approach—kill the cancer as the primary goal and hope for the best in terms of side effects. My mom beat the cancer, but the treatments left their mark, causing permanent neuropathy in her feet.

Today, generative AI offers personalized treatment plans, analyzing radiological images of individual tumors down to the genetic level. AI can analyze mammograms and ultrasounds with razor-sharp precision, catching things human eyes might miss. It enables oncologists to craft highly personalized treatment plans—targeted therapies that attack cancer cells while sparing as much healthy tissue as possible. Even better? Generative AI now has a lower error rate than human radiologists when interpreting mammograms and ultrasounds. Sure, both humans and AI can make mistakes—nobody's perfect—but studies show AI has a lower rate of false positives (saying there is cancer when there isn't) and false negatives (missing cancer when it's actually there). That's huge. Fewer false alarms, fewer missed diagnoses, and way more peace of mind to craft highly personalized treatment plans—targeted therapies that attack cancer cells while sparing as much healthy tissue as possible. My mom deserved that. Every patient deserves that.

Smart Tech, Smarter Healthcare

Hospitals today harness AI to predict patient influx, optimize resources, and even manage reimbursement more efficiently. Picture this: A healthcare system uses AI to sift through zettabytes of electronic health records, Medicare claims, diagnostic data, even EKG signals. It can analyze the locations of doctors' offices, patient ethnicity, income levels, age distributions, and more to understand the unique most common diseases and survival rates of specific regions. It evaluates the most common illnesses within a specified radius, air quality levels, prevalence and types of viral infections, bacterial infection types and rates, and the list goes on. AI can predict how many patients might flood the ER next week, the most common reasons, and which patients might face the toughest battles based on social determinants like income or geography—in addition to their medical health records.

And here's where it gets even cooler: AI doesn't just predict patient trends. It processes all this massive, messy data to help hospitals figure out exactly how many doctors they need to hire, what types of specialists they should bring in, how many surgeons (and which kind), how many ER beds they need, the number of nurses and office staff, and even how many ambulances they should have on standby. It generates recommendations that help health systems optimize the balance between high-quality care and the cost of delivering that care.

Now imagine that a hospital used generative AI to generate an optimal operating model. They'd have the right number of beds, the right equipment, and the right team of doctors, surgeons, nurses—you name it—to manage patient flow smoothly. In one real-world case, a hospital aimed to reduce ER visits from four to six hours down to just one hour. And they actually pulled it off. That's the kind of game-changing potential we're talking about. It's healthcare that's proactive rather than reactive.

Risk Management and Patient-Centered Care

Let's be honest—whether you get your insurance through the company you work for, purchasing it yourself, or you're on Medicare or Medicaid, dealing with insurance can be painful, frustrating, and downright maddening. We've all been there: A doctor orders a much-needed procedure, you hold your breath, and—bam—your insurance denies it. So, you work with your doctor to write appeals, fight back, and cross your fingers. Sometimes you win, sometimes you lose, and sometimes you get so worn down that you just give up, living with your medical condition and hoping next year's insurance plan will finally cover it.

And if you've ever had a procedure and looked at the bill, you've probably noticed some wild math: the "actual" cost of the procedure, then a mysterious "insurance adjustment" column, and then—miraculously—you only owe, say, $150. That magic? It's your insurance company negotiating what they'll actually pay the hospital or doctor, leaving you to pick up the rest. Depending on your plan, that could mean a few hundred bucks—or a few thousand. Healthcare finance is a jungle, my friends.

Hospitals today harness AI to predict patient influx, optimize resources, and even manage reimbursement more efficiently. Doctors, for example, want to minimize their risk of losing money and prefer contracting with the insurance companies on what is called a fee-for-service model. A fee-for-service model is a traditional way healthcare providers get paid. Insurance companies aren't exactly fans of this setup because they take on all the financial risk. If the services performed don't actually improve patient health, the insurance company still has to shell out the cash. Plus, fee-for-service can sometimes incentivize doctors to overuse the system—ordering extra tests, additional visits, or unnecessary procedures—because the more services they provide, the more they get paid, even if it's not always what's best for the patient.

Insurance companies, on the other hand, prefer the opposite end of the spectrum. In what's called a "global budget," the insurance company negotiates with the provider and they agree on a total budget for the year. They pay the provider that amount—and they're done. It's up to the provider to manage their operations within that fixed budget. If the provider runs a tight, efficient ship and stays under budget, they get to keep the profits. But if they run inefficiently, order too many tests, or see more patients than anticipated, they lose money. From the insurance company's perspective, it's a dream: they have virtually no financial risk of losing money.

Insurance contracts between providers and payers aren't simple one-size-fits-all deals—there are literally hundreds of different contracts, each with its own mix of risk, payment models, and performance incentives. And that's where AI comes in. Instead of manually reviewing endless combinations, haggling over terms, and going through a mind-numbing back-and-forth, generative AI can model the optimal structure based on how much risk both the insurance company and provider are willing to take. It can quickly generate an initial recommendation for the contract, letting both sides review and tweak it. Even better, if either party suggests changes, an AI agent can instantly rework the model and produce a new proposal. It dramatically speeds up the negotiation process—and probably lands everyone at a much smarter, more balanced risk-share model than traditional methods ever could. And hopefully your share out of your pocket is as small as possible.

Doctors Reimagined: How AI Is Helping Physicians Help Patients

Doctors themselves are also tapping into the power of generative AI in big ways. Medical education about new therapies coming to market is now being delivered to doctors in advance of approval, giving them a head start. With AI, a doctor can generate quick recommendations for the types of patients who would be appropriate candidates for new treatments. It can even suggest where a new therapy might fit into that doctor's existing treatment plans—something that used to take endless rounds of peer discussions, literature reviews, or waiting for a visit from a pharmaceutical company's medical science liaison. If they have questions, instead of tracking down a busy colleague or waiting days to hear back from pharma, they can ask an AI agent directly and get instant, evidence-based responses.

But that's just the start. AI can assist doctors in reviewing patient histories, flagging potential contraindications, and even suggesting alternative therapies if the first line of treatment doesn't work. It acts as a kind of supercharged physician assistant—not a replacement, but an augmentation. The doctor still makes the call, carefully reviewing every AI-generated recommendation and deciding whether to accept, modify, or ignore it based on their clinical judgment and what's best for each individual patient.

Patient Empowerment: Digital Natives Take Charge

Today's patients aren't waiting quietly for the doctor to tell them what's what. Thanks to smartphones, wearables, and AI-driven chatbots, patients are learning more, questioning more, and expecting more. Conversational AI lets people interact with healthcare quickly, intuitively, and empathetically. Instead of filling out cold, impersonal forms, patients can now have natural conversations with a virtual nurse who flags potential issues before they become emergencies. These AI agents can triage symptoms, recommend next steps, and schedule appointments. In many cases, they can identify red flags early—allowing real nurses and doctors to step in sooner and potentially prevent bigger problems down the line.

Patients are becoming more empowered than ever before, gathering high-quality medical advice to educate themselves and have more productive conversations with their doctors. Through a single AI-powered chat agent, they can perform research, identify products and therapies that may help treat their conditions, and even ask the AI to break down complex clinical trial data into plain, easy-to-understand language. Instead of feeling overwhelmed by medical jargon, patients can show up to appointments informed, confident, and ready to collaborate with their doctors on the best course of action. It's leveling the playing field and giving patients a stronger voice in their own healthcare journey.

Closing the Loop: Technology with a Heart

Generative AI isn't magic. It hallucinates. It gets things wrong. It can be biased. It needs oversight, just like any tool wielded by humans. But when we use it thoughtfully—integrating it from drug discovery to the final follow-up with a cancer survivor—everyone wins.

It's not about replacing doctors, nurses, or scientists; it's about giving them superpowers. It's about making healthcare faster, cheaper—and better. Behind every algorithmic recommendation, there's a real person, a real family, a real life waiting for a second chance. That's a future worth fighting for.

About the Author

Kurt Mueller has spent more than two decades at the forefront of pharmaceutical digital marketing, helping shape how life sciences brands embrace emerging technologies. As a founder and owner of several companies—including a software company he launched with his brother—Kurt brings deep entrepreneurial experience and a self-taught drive that's fueled both failure and major wins. He's earned a reputation as a trusted voice in the industry, contributing to respected publications and speaking at top conferences across the country. Known for blending business savvy with a love of innovation, Kurt is passionate about how generative AI and other digital tools can help healthcare professionals deliver smarter, faster, more personalized care.

Email: kurt.mueller@boundlesslife.com

LinkedIn: http://www.linkedin.com/in/kwmueller

CHAPTER 22

ENGINEERING THE FUTURE: AI PRIVACY AND THE RISE OF THE AGENT-DRIVEN ECONOMY

By Giorgio Natili
VP and Head of Engineering
Seattle, Washington

It is not enough for machines to be intelligent; we must ensure they are aligned with human values.

—Stuart Russel

In a traditional economy, humans are the primary decision-makers and actors. In contrast, the agent-driven economy envisions AI agents—autonomous software entities—as key participants in economic processes. These agents can handle tasks such as customer service, logistics, financial analysis, and even inter-agent negotiations.

A significant development in this realm is the agent-to-agent (A2A) economy, where AI agents interact and transact with each other, automating processes that were traditionally managed by humans.

An AI agent is an autonomous software entity designed to process contextual inputs to determine actions, process information, and take actions to achieve specific goals without continuous human intervention. These agents can operate in various environments—physical, digital, or hybrid—by utilizing sensors, data inputs, and interfaces to gather information, make decisions, and execute tasks.

AI agents are characterized by their ability to "perceive" (or evaluate) environmental signals, make decisions based on that "perception," and take actions to achieve specific goals. In digital contexts, "perception" involves processing data inputs such as user interactions, system logs, or external data sources. The agent interprets this information to understand the current state of its environment. Based on this understanding, the agent makes decisions using predefined rules, learned experiences, or predictive models. These decisions lead to actions, such as sending notifications, adjusting system settings, or initiating other processes. Advanced AI agents also possess learning capabilities, allowing them to adapt (and in some systems evolve) their behavior over time based on new data and outcomes of previous actions.

The integration of AI agents into various industries is transforming our workforce. While some roles may be displaced, this shift also presents opportunities for human workers to engage in more complex, creative, and strategic tasks. By automating routine and repetitive functions, AI agents can enhance productivity and efficiency, allowing humans to focus on areas that require emotional intelligence, critical thinking, and innovation.

The Algorithmic Tightrope: Balancing Promise and Peril in the Agent Economy

The agent-driven economy promises unprecedented efficiency and innovation, a fundamental change driven by autonomous AI. Yet, this rapid advance necessitates a critical compromise: Direct human control diminishes as agents independently manage critical processes and sensitive data. For companies and individuals navigating this new terrain, it's a challenging balancing act amid a landscape of new vulnerabilities. This predicament isn't merely technical; it demands a confrontation with the utility-privacy paradox—the conflict between the desire for personalized, convenient digital services (utility) and the need to protect personal data and privacy—and the amplified potential for data misuse, compelling a new calculus of trust and vigilance if we are to harness AI's power responsibly.

This new landscape is dominated by the insatiable agent, whose operational effectiveness and increasing power are directly proportional to the volume and sensitivity of the data it consumes. These AI entities pursue deeper specialization through continuous pre-training, fine-tuning, and learning—often consuming an organization's most sensitive data in the process. While major LLM providers (OpenAI, Google AI, Anthropic, and Cohere, among others) publicly outline privacy policies, offering data management controls and enterprise assurances against general model training (Source: Review of publicly available Open AI, Google AI, Anthropic, and Cohere privacy policies, 2023–2024), these policies can appear opaque when scrutinized against the lifecycle of an autonomous, learning agent. A critical question thus emerges: As agents interact and potentially assimilate confidential inputs, how can verifiable data confidentiality be guaranteed? The drive for hyper-specialized AI now targets precisely this sensitive data, escalating both the rewards and the risks tied to these powerful, yet data-hungry, algorithmic entities.

Many organizations rely upon established defenses—legal frameworks like GDPR and CCPA, coupled with technical security measures such as encryption at rest and in transit—that falter before these AI-driven challenges. Conceived for more static data paradigms, these methods function like outdated locks—no match for AI agents equipped with novel keys forged from inference and scale. Consider an AI agent optimizing a supply chain: It interacts with numerous vendor agents. During these A2A negotiations, subtle communication patterns or metadata analysis across the agent network can inadvertently reveal aggregated fragments of sensitive pricing or inventory data. Such pathways for potential data leakage bypass traditional perimeter security and static encryption protocols, which lack the design to monitor or prevent this dynamic, distributed interplay. Consequently, auditing the decision-making processes of these continuously learning agents—a cornerstone of accountability—becomes exponentially more complex than tracking human-driven actions, making older compliance frameworks ill-equipped for the agent-driven economy's scale and speed.

Failing to address these emergent privacy complexities risks more than non-compliance; it undermines long-term strategic viability. Organizations face significant financial penalties under evolving AI-specific regulations. Beyond fines, AI-driven ethical breaches can irrevocably damage brand reputation and erode stakeholder trust—a currency far more difficult to recover than capital. Users, increasingly conscious of data vulnerabilities, will hesitate to engage with systems whose data-handling practices lack transparency. Yet, a critical tension persists: The most potent AI models—especially in high-stakes domains like financial forecasting or medical diagnostics—require access to confidential data to reach peak accuracy. Proactively engineering privacy into these systems isn't just risk mitigation; it's a strategic necessity for earning trust and unlocking a lasting competitive edge through secure, data-driven insights.

Successfully navigating this algorithmic tightrope demands more than incremental adjustments; it requires a fundamental reimagining of privacy and security for the agent-driven era. Simply put, the old blueprints for data governance are insufficient. The vulnerabilities are novel, the scale unprecedented, and the stakes—encompassing financial stability, corporate reputation, and societal trust—are exceptionally high. Organizations must move beyond reactive compliance and actively champion the development and adoption of inherently private and secure AI systems. This proactive approach is not merely defensive; it is the foundation upon which sustainable innovation and enduring competitive advantage will be built in an economy increasingly shaped by autonomous intelligence.

Data Sovereignty: Reclaiming Agency in an AI World

As AI systems become more deeply embedded in every aspect of business and society, the question of who truly owns, controls, and benefits from data moves from a technical concern to a core strategic and ethical imperative. Data sovereignty means that the ultimate authority over data—whether personal, community-held, or organizational—should reside with its originators or rightful stewards. It represents a foundational shift toward empowering individuals, communities, and organizations to govern their digital destinies in an age of intelligent machines.

The assertion and realization of data sovereignty are critical for fostering individual autonomy, enabling community empowerment, ensuring organizational integrity, and, crucially, for developing the trustworthy, human-centric AI systems that will deliver sustainable economic and positive societal transformation. Without robust data sovereignty, the promise of AI risks being undermined by a crisis of trust and an erosion of control. What happens when current AI paradigms challenge this sovereignty—and what emerging technological, policy-driven, and strategic pathways might help us reclaim it?

Data sovereignty is the fundamental right and practical ability to exercise control over data. It transcends data privacy (which focuses on protection from unauthorized access) and data security (which concerns the integrity and availability of data), asserting a more profound authority over the entire data lifecycle. How data sovereignty takes shape across individuals, communities, and organizations carries profound implications for the AI era.

Individual Data Sovereignty

At its core, individual data sovereignty envisions a future where people possess tangible and actionable control over their personal information, especially as it fuels and interacts with artificial intelligence. This ideal, however, stands in stark contrast to the current reality. The ambition is for individuals to truly understand what personal data is being used and for what specific AI-driven purposes—be it for algorithmic profiling, personalized recommendations, or automated decision-making. But, such clarity is often not enough and it might be very elusive. An information asymmetry prevails, with many AI systems operating as impenetrable "black boxes."

The Cambridge Analytica scandal starkly illustrated this, where personal data harvested from millions of Facebook users was used to build psychological profiles for political targeting, far beyond the users' original understanding or consent for such inferential use. Individuals may be vaguely aware their data is in play, but the "how" and "why" of its processing, or the subtle inferences drawn from it—such as an AI discerning potential health conditions from seemingly innocuous shopping patterns—remain obscured, gutting the very notion of informed consent.

This opacity directly undermines the power to decide whether and how one's data is used by AI models. Instead of granular choice, people are typically confronted with take-it-or-leave-it terms of service, common across most social media platforms and free app services, where consent for a vast array of data uses, including extensive AI processing, is bundled as a non-negotiable prerequisite for accessing a service. The opportunity to selectively permit or deny specific AI applications of their data is seldom offered, and information initially collected for one purpose frequently finds new life training entirely different AI models, all without seeking fresh, explicit consent.

Even when rights like data access, rectification, or deletion are legally recognized, putting them into practice in the complex world of AI is extremely difficult. An individual might request access to their data, only to receive a report that is overwhelming, incomplete, or both. The ability to rectify inaccuracies is crucial, as flawed data can lead to biased or harmful AI outcomes. Yet "untraining" specific data points from a deeply embedded AI model—or correcting all its downstream inferential impacts—is a significant technical challenge. Likewise, deletion requests are often thwarted by the persistence of information in backups, derivative models, or distributed systems. People also have little control over who receives their AI-generated profiles or data-driven insights. A sprawling ecosystem of data brokers and tech companies facilitates the aggregation and resale of such information. For example, precise location data from seemingly harmless mobile apps is often collected and sold, enabling detailed tracking. As a result, tracing the journey of one's data—let alone reclaiming control—is nearly impossible.

Compounding these issues, prevailing privacy models often emphasize the initial collection of data rather than its subsequent use and downstream impact in AI systems. Companies can invoke "legitimate interests" to justify continued data processing for model development, frequently overriding individual preferences. These practices are rooted in dominant digital business models that depend on mass data exploitation, creating systemic resistance to any paradigm that would grant individuals genuine control. Until these foundational obstacles are addressed, individual data sovereignty remains more an aspiration than a reality. People are left as passive subjects rather than active participants in shaping their digital lives—a disempowerment with profound implications for personal autonomy and dignity in an increasingly AI-mediated world.

Community Data Sovereignty

Community data sovereignty builds on the individual right to data control by extending it to collective identities. It represents a shared aspiration for digital self-determination—especially among cultural groups whose languages, traditions, identities, heritage, and collective knowledge are increasingly digitized and thus newly vulnerable. At its core, community data sovereignty means empowering these groups to govern how their data is collected, interpreted, and used—particularly in AI systems. This includes protecting against exploitative practices, such as when Indigenous art styles are ingested by AI image generators to create derivative works without consent, credit, or compensation.

Yet realizing this vision faces formidable systemic barriers. For many communities, especially those shaped by colonization or marginalization, vast troves of data already reside in external institutions. This includes everything from ancestral remains and artifacts to ethnographic records and linguistic corpora, often gathered without consent and with little return of value to the communities themselves. Legal frameworks further compound the problem. Most data rights regimes focus on individuals, states, or corporations, leaving communities in a legal grey zone, unrecognized as collective rights-holders over their cultural and biological data.

Without clear mechanisms for sovereign control, communities risk seeing their stories distorted and their data used to train models that reinforce stereotypes or erase cultural nuances. These risks are heightened by the digital divide. Many minority language or under-resourced communities lack the data or infrastructure to develop culturally aligned AI, forcing dependence on generic tools built by large tech companies with different priorities. Addressing this imbalance requires more than resistance; it calls for building legal, technical, and financial pathways that empower communities to shape AI on their own terms. Only then can we move beyond digital colonialism toward inclusive, community-led innovation.

Organizational Data Sovereignty

In an AI-driven economy, organizational data sovereignty is a cornerstone of strategic autonomy and competitive resilience. It represents an enterprise's ability to govern its data and AI systems throughout their entire lifecycle—from data ingestion and model development to deployment, operational oversight, and eventual decommissioning. At stake is the ability to ensure that all AI activities align with the organization's core values, strategic objectives, intellectual property protections, and regulatory obligations—even when interfacing with external vendors and infrastructure. However, achieving this vision is increasingly difficult in today's interconnected ecosystem.

External dependencies post immediate challenges. Most organizations rely on cloud providers and AI-as-a-service (AIaaS) platforms that operate outside their control. Regulatory decisions such as Schrems II have complicated cross-border data flows, while laws like FISA 702 in the US raise fears that sensitive data may be accessed by foreign governments. Even when contractual safeguards exist, organizations face risks: Client data may inadvertently feed a vendor's foundational models or fall prey to provider-level security breaches. Once data enters a partner or supplier's system, maintaining oversight becomes a delicate balancing act.

Internal risks are no less serious. The rise of "shadow AI"—where employees use public tools like ChatGPT, Claude AI, or Gemini for sensitive tasks—has led to leaks of proprietary code, internal memos, and IP. Meanwhile organizations must navigate a fragmented and fast-changing regulatory landscape, from China's Cybersecurity Law to sector-specific AI standards, each carrying its own legal exposure.

Structural barriers further constrain progress. The cost of building sovereign infrastructure, such as dedicated private clouds or deploying privacy-enhancing technologies (PETs), can be prohibitive. There's also a widening skills gap: Few professionals combine AI expertise with governance, compliance, and operational security. And even when protections are in place, organizations must defend against a new class of risk: the exposure of proprietary insights via LLMs that memorize training data or reproduce sensitive material under sophisticated prompting attacks.

To move forward, enterprises must treat data sovereignty not as an aspirational principle but as a strategic priority. That means investing in auditable, policy-aware data pipelines, adopting PETs that ensure control without sacrificing performance, and cultivating talent equipped to manage the governance demands of sovereign AI. Only then can organizations unlock AI's value.

The drive for data sovereignty is not merely technical or legal—it is deeply ethical. As data becomes an extension of individual and collective identity, losing control over it undermines agency, erodes trust, and compromises strategic autonomy. Reclaiming data sovereignty is, at its core, about restoring that agency—ensuring the digital world empowers individuals, communities, and organizations alike.

The Pillars of Data Sovereignty

Asserting data sovereignty becomes paramount in the age of autonomous AI agents—systems that operate with increasing independence in collecting, processing, and acting on data. To manage these agents effectively, organizations must move beyond reactive mindsets and build proactive, principled frameworks for control. This shift is essential not only for safeguarding privacy but for ensuring that autonomous systems align with human values, organizational goals, and ethical standards. Operationalizing sovereignty at this level requires three foundational pillars: a comprehensive governance and policy framework, a resilient technical infrastructure with strong security guarantees, and effective data control and interoperability mechanisms. Together, these pillars form the bedrock for building AI systems that are not only strategically advantageous but also trustworthy stewards of human dignity—ensuring people can engage authentically, without fear of opaque judgment or misuse.

Governance and Policy: Defining Boundaries and Accountability

To uphold data sovereignty in systems governed by autonomous AI agents, organizations must begin with a comprehensive governance and policy framework. This pillar is not a set of static documents but a dynamic, organization-wide system that defines authority, assigns responsibilities, and codifies principles that guide all data and AI-related activities. Its purpose is to ensure that autonomous AI aligns with the organization's ethical standards, legal obligations, risk posture, and strategic priorities—particularly in how it handles and processes sensitive data.

Effective governance starts with clear accountability structures. Key roles—such as a chief data sovereignty officer or an AI governance board—should be formally empowered to oversee data stewardship, privacy compliance, AI ethics, and model risk management. Without unambiguous oversight, autonomous agents might operate beyond their intended scope, creating loopholes for unintended data uses or breaches of sovereignty.

Equally essential are lifecycle-wide policies that govern how data is sourced, processed, retained, and decommissioned. For autonomous agents, these policies must include constraints on permissible levels of autonomy, triggers for human intervention, transparency requirements (even when explainability is partial), and clear protocols for handling deviations from policy. Robust risk assessments, such as data protection impact assessments (DPIAs), alongside continuous monitoring and AI-specific audits, ensure that safeguards remain effective as technology evolves. Organizations can look to standards like the NIST AI Risk Management Framework, the requirements and principles outlined in the EU AI Act, and AI management system standards like ISO/IEC 42001 to guide the construction of this governance pillar.

The key takeaway: A strong governance and policy framework turns the principle of data sovereignty into enforceable organizational reality—ensuring agents act in service of human-defined goals, not outside them.

About the Author

Giorgio Natili is an engineer at heart, a teacher by passion, and a community builder by choice. With over two decades spent leading high-impact engineering teams across companies like Amazon, Mozilla, Capital One, and now Opaque Systems, Giorgio has seen technology evolve from simple scripts to sophisticated AI-driven ecosystems. But what's remained constant throughout his journey is his unwavering belief in people—whether that means empowering his teams, mentoring startups at MIT, or helping individuals navigate the ethical complexities of AI.

Giorgio's path into tech started with an insatiable curiosity, the kind that made him take apart gadgets just to understand how they worked. That same curiosity led him from Rome to the global tech stage, shaping innovations in browser privacy, digital payments,

Kindle rendering, and AI-powered customer experiences. His work has helped millions of people—often quietly, behind the scenes—by making technology faster, fairer, and more secure.

Despite the titles and patents, Giorgio remains grounded in his core mission: building technology that serves humans—not the other way around. As a part-time lecturer and advisor at Tacoma Community College, he thrives on learning from others just as much as teaching. He's especially passionate about ethical AI, data sovereignty, and making sure that the next generation of technologists inherit not just powerful tools, but the wisdom to wield them with care.

This chapter, like all of Giorgio's work, is an invitation: to think critically, build responsibly, and most importantly, to never stop learning.

LinkedIn: https://www.linkedin.com/in/giorgionatili/

CHAPTER 23

SCALING INNOVATION: AI, CLOUD, AND THE FUTURE OF BUSINESS GROWTH

By Lech Nowak
Forward-Thinking AI Engineering Leader
Wroclaw, Poland

AI is the first technology in history which is not a tool, it's an agent. It could actually make decisions by itself. It can invent new ideas by itself.

—Yuval Noah Harari, *The Economic Times*, 2024

Introduction: AI Across the Business Lifecycle

As we navigate the Fourth Industrial Revolution, artificial intelligence has evolved from an aspirational technology to an essential business cornerstone. What began as experimental pilots in large enterprises has now democratized into accessible, transformative tools that organizations of every size can—and must—leverage to remain competitive. The businesses that will thrive are those recognizing that AI is no longer confined to specialized departments but functions as an omnipresent copilot throughout the entire business lifecycle.

The integration of AI across every business function represents more than technological adoption—it signifies a fundamental paradigm shift in how we conceptualize organizational growth. We are witnessing the convergence of sophisticated AI systems, enterprise-ready cloud infrastructures, autonomous agentic frameworks, vertical AI agents, no-code/low-code platforms, and seamless integration into existing workflows. This convergence is redefining what's possible for forward-thinking organizations willing to embed AI-first thinking into their operational DNA.

This chapter explores how modern AI technologies transform each phase of the business lifecycle, providing a roadmap for leaders seeking to harness these capabilities for sustainable growth. We examine how advanced prompting strategies create a new language for cross-functional innovation, how agentic AI systems enable autonomous workflows, and how cloud-native AI ecosystems provide the infrastructure needed to scale these solutions enterprise-wide.

Ideation and Strategy: AI-Enhanced Vision

The genesis of business growth begins with ideation—the process of conceptualizing opportunities that align with market needs. AI has transformed this traditionally human-centered activity into an augmented intelligence exercise, where human creativity meets computational pattern recognition.

Modern strategic planning now leverages tools like OpenAI's GPT-4.5 and o3-pro, as well as Anthropic's Claude 4 to analyze market signals across unprecedented volumes of data. These models can process and synthesize information from diverse sources—from industry reports to social media sentiment—identifying emerging trends before they become obvious to competitors. The critical advantage comes from the velocity and breadth of pattern recognition that AI provides, enabling strategic foresight that was previously impossible.

Specialized no-code platforms like Dust.tt and Relevance AI have transformed strategic intelligence gathering. These platforms enable business strategists without technical backgrounds to create sophisticated AI workflows that continuously monitor market signals, identify emerging opportunities, and generate actionable competitive insights. By visually designing AI agents that can analyze everything from market research to patent filings to regulatory changes, strategic planners can maintain continuous environmental awareness without requiring dedicated data science resources.

The rise of vertical AI agents—domain-specific assistants tailored for particular industries—has further transformed strategic planning. These specialized agents embed industry-specific knowledge, regulatory frameworks, and strategic patterns, enabling more nuanced analysis than general-purpose AI systems. Financial services firms leverage vertical agents that understand market mechanics, regulatory constraints, and historical patterns specific to their sector, enabling more precise opportunity identification and risk assessment.

For small and medium enterprises, the democratization of advanced AI capabilities through open-source models has dramatically leveled the playing field. A boutique consulting firm can now leverage sophisticated AI to develop market insights that previously required teams of analysts at larger competitors. This accessibility enables David-versus-Goliath scenarios where agility and AI adoption can overcome traditional resource advantages.

Product Development: From Conception to Creation

The product development lifecycle has been radically compressed through AI-augmented workflows. What once required months of research, design, prototyping, and testing can now unfold in weeks or days through intelligent automation and generative design.

The integration of AI into design tools like Figma has transformed the product development process from conception to implementation. Figma's AI-powered features enable designers to generate design elements, automate repetitive tasks, and enhance prototyping workflows with unprecedented efficiency. More significantly, integrations with platforms like Builder.io and Lovable have created a seamless pipeline from design to deployment. Designers can now create interfaces in Figma, use Builder.io to convert these designs into code, and leverage Lovable to add backend functionality and deploy full-stack applications—all with minimal traditional coding.

Developer-centric platforms like Cursor and Replit Agent have created AI-augmented environments that dramatically accelerate ideation and implementation. These intelligent coding environments combine the power of large language models with specialized development tools, enabling developers to translate conceptual ideas into functional applications with unprecedented speed. This evolution moves beyond simple code completion to true AI-developer partnership, where systems understand architectural patterns, implementation best practices, and performance considerations.

Multi-agent frameworks like CrewAI and Microsoft's Semantic Kernel enable product teams to orchestrate specialized AI agents across the development process. A product manager can define high-level requirements that are automatically translated into technical specifications by one agent, while another generates design mockups, and a third evaluates user experience implications. These agentic AI systems are increasingly capable of autonomous decision-making within their domains, handling complex development tasks with minimal human oversight while adhering to predefined quality standards and architectural principles.

Platforms like Vercel v0 and Bolt.new enable the rapid translation of conceptual ideas into functional applications. These tools represent a paradigm shift in the development workflow—allowing product teams to move from natural language descriptions directly to working prototypes without intermediate steps. This approach dramatically accelerates the concept-to-validation cycle, enabling product teams to test more ideas with users before committing to full implementation.

For software products, GitHub Copilot, Amazon Q Developer, and Claude Code have transformed development workflows. These tools go beyond code completion to offer architectural recommendations, identify potential performance bottlenecks, and even suggest refactoring approaches to improve maintainability. The most sophisticated organizations are using these capabilities to build "AI-native" products—applications designed from the ground up to leverage and continuously improve through AI capabilities.

Marketing and Sales: Precision Engagement at Scale

Perhaps no business function has been more visibly transformed by AI than marketing and sales. The promise of "right message, right person, right time" has finally become achievable through the convergence of predictive analytics, content generation, and conversational AI.

No-code AI platforms like Botsonic, Landbot, AppyPie, FlowHunt, and Zapier/Make/n8n have transformed how marketing teams design and deploy customer engagement strategies. These intuitive platforms enable marketers to create sophisticated, personalized customer journeys without requiring programming expertise. Botsonic and Landbot's ability to generate customized, multichannel chatbots has revolutionized lead qualification and nurturing, while platforms like FlowHunt, AppyPie, and Zapier/Make/n8n empower marketing teams to visually design complex content personalization and workflow rules based on customer behaviors and preferences.

Advanced language models have revolutionized content creation. Marketing teams can now generate personalized content at scale, with AI systems adapting messaging to specific audience segments while maintaining brand voice and strategic objectives. This capability extends beyond simple copy variations to encompass entire campaigns, from email sequences to social media calendars to landing page copy.

The frontier of marketing AI lies in dynamic content optimization. Tools built on platforms like LangChain can continuously test messaging variations, analyze performance data, and autonomously refine content strategies. This creates "self-optimizing" campaigns that become more effective over time without requiring constant human intervention.

Sales processes have been transformed through AI-powered sales assistants. These systems can handle the complete spectrum of sales activities from lead qualification to negotiations, with unprecedented capabilities for understanding prospect needs and adapting pitches accordingly. The most advanced assistants leverage emotional intelligence capabilities to detect subtle signals in prospect communications, adjusting their approach based on detected interest, hesitation, or confusion.

The most advanced organizations are now implementing full-funnel AI orchestration, where marketing and sales activities are coordinated through intelligent systems that adapt messaging, channel selection, and engagement timing based on prospect behavior and feedback. This creates coherent buyer journeys that feel personalized even when operating at massive scale.

Operations and Supply Chain: Autonomous Optimization

Operational excellence has traditionally required painstaking process design, constant monitoring, and incremental improvement. AI has fundamentally altered this paradigm by enabling autonomous optimization—systems that continuously analyze performance data, identify inefficiencies, and implement improvements without requiring constant human oversight.

No-code AI workflow platforms like Relevance AI and Kore.ai enable operations specialists to build sophisticated process automation without requiring programming expertise. These platforms provide visual interfaces for designing AI-powered workflows that can handle complex operational tasks—from inventory optimization to maintenance scheduling to quality control monitoring. By combining drag-and-drop interfaces with powerful AI capabilities, these tools have democratized process automation across organizations, enabling frontline operations teams to identify and implement efficiency improvements without requiring IT involvement.

Edge AI deployments across supply chain operations represent another transformative approach. By implementing AI capabilities directly at operational endpoints—warehouses, vehicles, and manufacturing facilities—organizations can make critical decisions in milliseconds without requiring cloud connectivity. This approach has proven particularly valuable in logistics operations where network connectivity may be unreliable and where real-time adjustments to routing and scheduling can yield significant efficiency gains.

Manufacturing operations have similarly evolved through predictive maintenance and quality control. Computer vision systems can detect product defects with greater accuracy than human inspectors, while sensor networks combined with machine learning models predict equipment failures before they occur. These capabilities transform maintenance from a reactive to a predictive discipline, dramatically reducing downtime and extending asset lifespans.

The frontiers of operational AI lie in autonomous decision-making frameworks. Systems built on platforms like AutoGPT and CrewAI can manage routine operational decisions within defined parameters, escalating only exceptional cases that require human judgment. This approach creates highly scalable operations that maintain quality without proportionally increasing management overhead.

Customer Service and Support: Intelligent Assistance

Customer experience has become the primary competitive battleground across industries, and AI-powered support systems have emerged as critical differentiators. The evolution from simple chatbots to conversational AI represents a quantum leap in capability, with modern systems able to understand context, maintain coherent dialogue, and resolve complex issues without human intervention.

The widespread adoption of multimodal AI models in customer support has transformed service delivery. These systems can seamlessly process and respond to multiple information formats—text, voice, images, and video—creating support experiences that mimic natural human interaction. A customer can show a malfunctioning product through their smartphone camera, explain the issue verbally, and receive both visual and spoken guidance in response.

The democratization of conversational AI through no-code platforms has transformed how organizations design and deploy customer support solutions. These platforms enable support teams to create sophisticated AI assistants without requiring programming expertise. By embedding domain-specific knowledge and integrating with existing support systems, these platforms enable even smaller organizations to offer 24/7 intelligent assistance that previously required extensive technical resources.

Knowledge management—long the Achilles' heel of customer support—has been revolutionized through retrieval-augmented generation (RAG) approaches. Systems built on frameworks like LangChain can instantly access and synthesize information from product documentation, support histories, and even internal communications to provide accurate, contextual answers. This capability ensures consistency across support channels while reducing the training burden for human agents.

Proactive support represents the frontier of customer experience AI. Systems can analyze usage patterns, identify potential frustration points, and proactively offer assistance before customers need to request help. This approach transforms support from a reactive cost center to a proactive value driver that enhances customer satisfaction and retention.

Maintenance and Continuous Improvement: Learning Systems

The traditional approach to product and service improvement relied heavily on formalized feedback channels and scheduled update cycles.

AI has transformed this paradigm by enabling continuous learning from every customer interaction, creating products and services that improve organically through usage.

Autonomous experimentation frameworks can independently design, implement, and evaluate product and service variations without requiring continuous human oversight. By systematically exploring solution spaces through carefully controlled experiments, these frameworks dramatically accelerate the pace of innovation and optimization.

Developer-centric platforms have transformed how organizations implement continuous improvement for digital products. These AI-augmented development environments enable seamless transitions from feedback to implementation, with systems capable of understanding user issues, generating appropriate fixes, and deploying updates with minimal human involvement.

Modern improvement frameworks leverage sentiment analysis and natural language processing to extract actionable insights from unstructured feedback sources—support conversations, social media mentions, review sites, and community forums. Systems can identify emerging issues, prioritize enhancement opportunities, and even suggest specific improvements based on aggregated customer experiences. Software products have embraced "AI-native" architectures that continuously optimize user experiences. These systems analyze usage patterns, identify friction points, and automatically adapt interfaces and workflows to better serve individual users. The result is personalized experiences that become more intuitive over time without requiring explicit configuration.

Prompt Engineering: The New Language of Business Innovation

As AI capabilities have expanded, prompt engineering has emerged as a critical discipline for unlocking their full potential. What began as simple text inputs has evolved into sophisticated frameworks that enable precise control over AI outputs and behaviors.

Business Users and Prompt-Based Collaboration

No-code AI platforms have revolutionized prompt engineering for business users. These intuitive platforms enable non-technical team members to create sophisticated AI workflows through visual interfaces, dramatically expanding the pool of potential AI creators within organizations. Business analysts, marketers, and operations specialists can now design complex prompt sequences, integrate multiple data sources, and implement decision logic without writing code.

Chain-of-thought prompting represents a particularly powerful technique for business applications. By guiding AI systems through explicit reasoning steps, users can ensure that outputs reflect appropriate analytical processes rather than simple pattern matching. This approach is especially valuable for financial analysis, strategic planning, and other domains where process transparency is essential.

Role-based prompting has similarly transformed how business users interact with AI systems. By establishing clear context ("You are a marketing strategist with expertise in B2B software") and defining evaluation criteria, users can elicit more relevant and actionable outputs. This technique enables even casual users to access specialized AI capabilities without requiring technical configuration.

The frontier of business prompting lies in collaborative frameworks where multiple stakeholders can refine and evolve prompts over time. Platforms enable teams to create, share, and iteratively improve prompts as organizational assets, creating institutional knowledge around effective AI interaction patterns.

Agentic AI Systems: Autonomous Workflows

The emergence of agentic AI frameworks represents perhaps the most significant advancement for business transformation. These systems go beyond responding to specific prompts to actively pursuing goals with minimal human oversight. Platforms like AutoGPT, BabyAGI, and CrewAI enable organizations to define high-level objectives and delegate their execution to autonomous AI agents that can plan, reason, and adapt to changing conditions. These frameworks represent a fundamental shift from AI as a tool to AI as a collaborator capable of independent problem-solving.

Multi-agent architectures take this capability further by orchestrating teams of specialized AI agents working in concert. A typical workflow might involve a planning agent breaking a problem into subtasks, research agents gathering relevant information, analysis agents processing the data, and implementation agents executing the required actions—all coordinated through a central orchestration layer.

The most advanced implementations leverage agent memory and continuous learning. These systems not only execute tasks but also store experiences, learn from outcomes, and refine their approaches over time. This creates a virtuous cycle where agents become increasingly effective through usage, developing domain expertise and organizational context that enhances their decision-making capabilities.

For organizations implementing agentic AI, success depends on thoughtful integration with human workflows. The most effective approaches define clear boundaries for agent autonomy, establish robust oversight mechanisms, and create seamless handoffs between human and AI contributors. This collaborative model preserves human judgment for critical decisions while leveraging AI for routine tasks and decision support.

Conclusion: The AI-First Imperative

The transformation outlined in this chapter represents more than technological evolution—it signifies a fundamental reimagining of how businesses operate and compete. Organizations that embrace AI-first thinking across their operations gain unprecedented advantages in agility, efficiency, and customer experience.

For leaders navigating this transformation, success depends on three key principles:

1. *Democratization:* making AI capabilities accessible throughout the organization through no-code platforms, embedded tools, and intuitive interfaces
2. *Integration:* weaving AI seamlessly into existing workflows and systems rather than treating it as a separate technological silo
3. *Augmentation:* designing AI implementations that enhance rather than replace human capabilities, creating partnerships that leverage the unique strengths of both

The businesses that will thrive in this new era will be those that recognize AI not as a technological investment but as a strategic imperative that touches every aspect of their operations. By embedding AI-first thinking into their operational DNA, forward-thinking organizations can transcend traditional limitations of scale, insight, and execution—unlocking new frontiers of innovation and growth.

About the Author

Lech Nowak represents the vanguard of AI-first thinking in enterprise technology transformation. For over two decades, he has navigated the convergence of cloud infrastructure, orchestrating end-to-end implementations that transcend traditional operational constraints across the telecommunications, travel, and automotive industries.

His technical expertise spans the full spectrum of modern AI capabilities—from multi-agent frameworks and search-augmented generation to cloud-native ML pipelines and autonomous workflow orchestration. He has championed the democratization of advanced transformation through microservices, no-code platforms, and intuitive interfaces, transforming distributed systems from niche technologies into ubiquitous business copilots.

Beyond technical execution, Lech focuses on human-AI partnership models that augment rather than replace human capabilities. He consistently embeds AI-first thinking into the DNA of organizations, guiding them from experimental pilots to fully integrated intelligent operations—ultimately delivering sustainable competitive advantage.

Email: lechnowak@lechnowak.com

LinkedIn: www.linkedin.com/in/lech-nowak

CHAPTER 24

AI: DATA QUICKSAND THREATENS THE FUTURE OF LAW

By Rory O'Keeffe
Lawyer, RMOK Legal
London, England, United Kingdom

Ní dhéanann droch-fhréamhacha crann maith.

(Bad roots don't make a good tree.)

—Irish saying

Imagine this: The courtroom was silent as the AI-powered legal team presented its closing arguments. Their analysis of millions of case files, executed in mere seconds, had revealed a crucial precedent that swayed the TikTok-generation jury. The opposing counsel, a small firm relying on traditional research methods, could only watch in awe—and a growing sense of unease. This wasn't just a victory; it was a glimpse into a future where legal success might be determined not by the strength of one's arguments, but by the speed of the AI tool used and the size of its dataset.

The rise of AI (though around since the 1950s), with its newly polished promise of unprecedented efficiency and insight, is like a gold rush or skyscraper growing higher and higher. But that skyscraper, I argue, is being built on a foundation of data quicksand, and the stability of the entire legal profession may be at risk. Let me be clear, I am and always will be the tech geek who enjoys TV shows and movies like *Tomorrow's World*, *Star Trek*, *Star Wars*, and *Dr. Who*. The potential of the new legal tech flooding the market is exciting, but we cannot ignore the need to apply legal and ethical scrutiny when assessing these "magical" tools. "Invest with care" has been the mantra for the last few years around all new technology (remember the Metaverse ... anyone?!?!).

From the advent of email, which streamlined document transfer, to the rise of e-discovery software and contract review tools, the legal industry has consistently sought technological solutions to enhance efficiency and reduce costs. Each innovation has been an attempt to build a more efficient "structure," a better way of organizing and delivering legal services. Now, "new" or agentic AI represents the most ambitious undertaking yet: systems that can not only process information but also autonomously make decisions and take actions. This is the tallest, most complex "structure" we've ever conceived. Yet, its reliance on massive datasets raises a fundamental question: What happens when that data is incomplete, biased, or inaccessible to all?

My view: AI continues to hold immense promise for the legal profession, but its reliance on vast, high-quality data creates a significant risk of exacerbating the gap between large, well-resourced firms and smaller practices; between big corporate and startups/scaleups. There are consumer and competition law questions there too. The vendors may have large databases to help smaller entities, but there remains the risk that you become overly reliant, linked, or locked into those arrangements. Vendors are collaborating with larger law firms to develop go-to-market AI legal software, which begs the question of whether a smaller firm will be forced to only be a "fast follower" or even always at a competitive disadvantage or worse, squeezed out of key legal fields. The Hollywood "underdog" courtroom drama might not be as easy to write in the future. We are facing a potential future where only the "data-rich" can effectively compete, threatening the very foundations of a fair and equitable legal system.

Navigating the Regulatory Maze (UK, EU, Global)

The regulatory landscape surrounding AI is as complex and varied as the technology itself. In the UK, a principles-based approach seeks to balance innovation with risk management. It remains to be seen if a version of Lord Holmes' Artificial Intelligence (Regulation) Bill will appear again. Though, the UK is being influenced by the EU's Artificial Intelligence Act, which attempts a more structured framework, categorising AI systems by risk—the act has received very mixed views and not the gold standard the EU suggested for the EU GDPR. When you also consider the American patchwork of federal and state regulations, with California's focus on data privacy, it adds another layer of complexity; and then all countries like Brazil, Canada, India, Australia, Singapore, Japan, UAE, and more are introducing AI action plans, policies, and legislation across many fronts. While these jurisdictions grapple with issues like transparency and accountability, they often overlook a more insidious problem: the uneven distribution of the data that fuels these systems. Do these regulatory approaches inadvertently favour large firms with the resources to invest in data acquisition and compliance, potentially solidifying their dominance? The answer, I fear, is a resounding "yes."

Key Legal Risks of AI Today: The Data Quicksand Beneath Our Feet

The legal risks associated with AI are well-documented: data privacy concerns, algorithmic bias, intellectual property dilemmas, and the potential for errors, hallucinations or malfunctions. There is not enough time to explore how the use of Gen AI tools for coding software has the potential (in theory) for coding malicious backdoors without the software developer and the buyer's knowledge. However, these risks are not isolated; they are interconnected and, crucially, exacerbated by the "data quicksand" upon which AI relies.

- *Data privacy*: The need for massive datasets to train effective AI systems often clashes with stringent data privacy regulations like GDPR and CCPA. Compliance becomes a significant burden, one that smaller firms, with their limited resources, may struggle to meet. This creates a two-tiered system, where large firms can leverage AI while smaller ones are hampered by compliance costs. Consider the complexities of GDPR, for instance. Its emphasis on data minimisation and purpose limitation requires firms to meticulously track the origin, use, and storage of every piece of data used to train an AI. For a multinational firm with dedicated compliance officers and substantial budgets, this is challenging but manageable. For a small firm, however, the costs of implementing such systems—hiring specialised personnel, investing in new software, and potentially restructuring data management practices—can be prohibitive. The same applies to the California Consumer Privacy Act (CCPA), with its focus on consumer rights and data transparency. The need to provide detailed disclosures about data collection and usage, and to respond to individual consumer requests places a significant administrative burden on legal practices. Again, larger firms with economies of scale are better equipped to handle these requirements, while smaller firms may find themselves struggling to keep up, potentially facing penalties or being forced to forgo the use of valuable AI tools.
- *Algorithmic bias*: AI systems are only as unbiased as the data they are trained on. If that data reflects existing societal biases, the AI will perpetuate and amplify those inequalities. Remember originally some GenAI models were only relevant up to 2023 and prohibited review of search engines. Smaller firms, lacking the resources to

curate diverse and representative datasets, are particularly vulnerable to this risk, potentially harming their clients and undermining the fairness of the legal process. Imagine an AI system trained on a dataset primarily composed of case law from a single jurisdiction with a history of discriminatory sentencing practices. Such a system, when used to predict case outcomes or suggest sentencing guidelines, would inevitably perpetuate those biases, regardless of the good intentions of the legal professionals using it. For a small firm handling a discrimination case, the inability to afford or access a more comprehensive and unbiased dataset could mean that their AI tools are actively working against their client's interests, potentially leading to an unfair outcome. Moreover, the difficulty in detecting and mitigating algorithmic bias further exacerbates this problem. Many AI systems, particularly the most advanced ones, operate as "black boxes," making it challenging to understand how they arrive at their conclusions. Responsible AI, or rather explainable AI, becomes paramount. Otherwise, this lack of transparency makes it difficult to identify the source of the bias and to correct it, particularly for firms with limited technical expertise.

- *Intellectual property*: The ownership and use of data in AI development raise complex IP issues. Firms that can afford to invest in proprietary data gain a significant advantage, creating a divide where access to cutting-edge AI tools is determined by financial resources, not legal expertise. Consider the scenario where a large firm develops a proprietary AI system for analysing patent infringement claims. This system is trained on a vast database of patents, case law, and technical publications, all meticulously collected and curated at considerable expense. The firm then uses this system to provide highly accurate and efficient patent litigation services, attracting more clients and further solidifying its market position. Smaller firms, lacking the resources to develop or acquire such a system, are left to rely on less sophisticated tools or manual methods, making it difficult for them to compete in this specialised area of law. Furthermore, the ownership of the AI-generated insights themselves raises thorny IP questions. If an AI system develops a novel legal argument or identifies a previously unknown precedent, who owns that intellectual property—the firm that developed the AI, the lawyers who used it, or the AI itself? These questions have significant implications for the future of legal practice and could further entrench the advantages of data-rich firms. Though, thankfully, on the IP case law front, there are a lot of cases around the world that have been exploring these legal challenges.

- *Errors and malfunctions*: Many AI systems operate as "black boxes," making it difficult to understand how they arrive at their decisions. When errors occur, determining liability becomes a complex and costly endeavour. Smaller firms, perhaps relying on less robust or well-tested AI tools due to budget constraints, face a disproportionately higher risk in such situations. For instance, imagine a small firm using an AI-powered contract review tool to expedite due diligence in a merger and acquisition deal. If the AI system, due to a flaw in its programming or inadequate training data, overlooks a crucial clause that exposes the client to significant financial risk, who is liable? Is it the firm that developed the AI, which may be a large corporation with substantial legal resources? Is it the small firm that relied on the tool, which may lack the financial capacity to defend itself in a protracted legal battle? Or is it the client, who ultimately suffered the consequences of the error? The lack of transparency in AI systems makes it incredibly difficult to establish a clear chain of causation and to allocate liability fairly. This uncertainty creates a chilling effect, particularly for smaller firms that may be hesitant to adopt AI tools if they fear being held responsible for errors they could not have prevented or detected. This continues to reaffirm my view that a human is always going to be needed, and the use of any AI tool will be just that, a tool enhancing the human using it.

AI in the Wild: More Than Just Chatting

AI systems are no longer confined to the realm of chatbots and simple automation. They can now establish objectives, formulate strategies, and execute tasks with minimal human intervention. From contract negotiation to legal research, these systems can perform complex functions, driven by the vast amounts of data they process. But this is where the "skyscraper" analogy becomes particularly apt.

These systems can reach incredible heights of efficiency and capability, but only if their data foundation is solid and evenly distributed. If not, they risk becoming unstable, unreliable, and ultimately, a liability. The ability of an AI to, for example, conduct complex legal research, hinges on its access to a comprehensive and constantly updated database of case law, statutes, and regulatory information. A large firm can afford to subscribe to multiple premium legal databases, ensuring their AI has access to the most current and reliable information. A smaller firm, however, may be limited to less expensive or open-source databases, which may be incomplete, outdated, or of questionable accuracy. This disparity in data access directly translates into a disparity in the AI's performance: The AI used by the large firm can conduct more thorough and accurate research, providing its lawyers with a significant advantage in case preparation and strategy development.

Real-World Shenanigans (And Hopefully Some Actual Benefits)

The allure of AI is undeniable, especially the newly (and heavily marketed) agentic AI. In contract law, AI can automate the drafting and analysis of agreements, potentially saving countless hours of lawyer time. In litigation, AI can sift through mountains of evidence to identify key patterns and insights. But as we explore these potential benefits, we must remain acutely aware of the data divide.

- *Accelerated contracting*: The promise of faster contract negotiation timelines—your "speed to close"—rests on the accuracy and completeness of the data used. For smaller firms, limited access to comprehensive, up-to-date legal databases could mean their information is less reliable or effective, hindering their ability to compete in increasingly automated transactions. Consider a scenario where a small firm is advising a client on a complex commercial transaction involving smart contracts. The AI system used by the firm needs to be able to access and interpret a vast array of legal and commercial data, including contract templates, regulatory filings, and industry-specific standards. If the firm's AI is trained on incomplete or outdated data, it may fail to identify crucial legal risks or to ensure that the contract complies with all applicable regulations. This could expose the firm and its client to significant financial losses and reputational damage, making them less competitive in the market for sophisticated legal services.
- *Global supply chains:* AI-driven supply chain management can optimise logistics and reduce risks, but smaller legal practices may lack the resources to access the vast datasets needed to train these systems effectively. This could leave them struggling to advise clients on complex international trade issues. In the realm of global supply chains, AI can be used to track the movement of goods, to monitor compliance with international trade regulations, and to predict potential disruptions. However, to do this effectively, the AI needs access to a massive amount of data, including shipping records, customs documents, and geopolitical information. Large multinational firms can invest in sophisticated data analytics platforms that provide them with this information in real time. Smaller firms, however, may have to rely on fragmented and less

reliable data sources, making it difficult for them to provide their clients with the same level of sophisticated advice. This could put them at a disadvantage when competing for clients involved in complex cross-border transactions.

- *Personalised marketing*: AI can enhance client acquisition and retention through personalised marketing, but smaller firms may find it challenging to gather and analyse the extensive data required to compete with the sophisticated marketing strategies of larger firms. The use of AI in legal marketing is becoming increasingly prevalent. AI-powered systems can analyse client data to identify potential leads, to personalise marketing messages, and to track the effectiveness of marketing campaigns. However, this requires access to a significant amount of data about potential and existing clients, including their demographics, online behaviour, and past interactions with the firm. Large firms with extensive client bases and sophisticated CRM systems have a clear advantage in this area. They can use AI to fine-tune their marketing efforts, to target specific client segments, and to maximise their return on investment. Smaller firms, with more limited client data, may find it difficult to compete with this level of sophistication, potentially hindering their ability to attract new clients and grow their practices.

The Looming Threat: A Two-Tiered Legal System?

The uneven distribution of data, coupled with the increasing reliance on AI, poses a grave threat: the emergence of a two-tiered legal system. In this scenario, large, data-rich firms wield AI with unparalleled effectiveness, leaving smaller firms struggling to compete. The consequences of such a divide are far-reaching. Access to justice could be compromised, as clients may be forced to seek representation from firms with the most advanced (and expensive) AI tools, which may still be cost-prohibitive law firms for many clients. Diversity within the legal profession could suffer, as smaller firms, which often serve diverse communities and provide opportunities for underrepresented lawyers, become increasingly marginalised. The overall health and competitiveness of the legal ecosystem would be diminished, replaced by a system where a few data-rich giants dominate the landscape.

Large corporate law firms are investing heavily in AI research and development, hiring data scientists and engineers, and building proprietary AI systems tailored to their specific needs. These firms are able to offer their clients faster, more efficient, and more sophisticated legal services, justifying higher fees and attracting even more high-value clients. Smaller firms, lacking the resources to make similar investments, are increasingly being relegated to handling less complex and less lucrative cases. This trend is not limited to large metropolitan areas; it is also affecting rural communities and smaller towns, where access to legal services is already a challenge. As AI becomes more integrated into the practice of law, the gap between the data-rich and the data-poor is likely to widen, creating a system where the quality of legal representation a client receives is increasingly determined by their ability to pay for access to the best AI-powered tools.

But there are potential solutions. We can promote data sharing and the development of open-source AI tools, ensuring that access to these technologies is not solely determined by financial resources. We can establish industry standards for data quality and transparency, fostering trust and ensuring that all firms, regardless of size, can rely on the data they use. We can provide training and resources to help smaller firms adopt AI responsibly, empowering them to leverage these technologies without being overwhelmed by the associated costs and complexities. And we can advocate for policies that address the data divide, recognising it as a systemic issue that requires a collective response.

One promising approach is the development of data cooperatives, where law firms, particularly smaller ones, could pool their resources and data to create shared datasets that can be used to train AI systems. This would allow them to achieve economies of scale and to compete more effectively with larger firms. Another solution is the creation of publicly funded research initiatives aimed at developing open-source AI tools for the legal profession. These tools could be made available to all firms, regardless of size, ensuring that everyone has access to the benefits of AI.

Conclusion: Building on Solid Ground

As the old Irish proverb reminds us, "*Ní dhéanann droch-fhréamhacha crann maith*"—bad roots don't make a good tree. AI stands at a crossroads. It offers the potential to transform the legal profession for the better, enhancing efficiency, expanding access to justice, and driving innovation. But this potential can only be realised if we address the fundamental challenge of data. We must ensure that the AI "tree" or "skyscraper" is built on solid ground, with strong, healthy roots in the form of reliable, accessible, and equitable data. This requires a collaborative effort involving legal professionals, technologists, policymakers, and academics. We must work together to cultivate a future where AI empowers the entire legal profession, not just a privileged few. The future of law, and the very principles of justice, may depend on it.

About the Author

Distinguished by thorough industry knowledge and a progressive mindset, Rory O'Keeffe founded his own practice, RMOK Legal. Through his work as a fractional general counsel, Rory has built a reputation for delivering incisive legal solutions to a diverse selection of clients, from fledgling startups to global enterprises. Leveraging cutting-edge tools, including automation platforms and AI, he stands at the forefront of innovation in commercial and technology law, providing services with expert precision, dedication, and a genuine client-first approach. His motto—*We look after it*—encapsulates RMOK Legal's philosophy.

Born in the historic, Celtic Ireland, now a thriving London-lawyer, he brings over two decades of experience across global technology and business transformation deals. Rory adeptly melds private practice, in-house and practical expertise with clear understanding of AI, cybersecurity, and emerging technologies. Previously, he worked as partner for an international firm and as a director of legal services in a Fortune 500 company. Rory has also launched a podcast covering tech and law, including all things AI, called *Beyond The Fine Print™*. Beyond legal work, Rory also offers a genuine commitment to the broader legal and social community, serving as a charity trustee at The Solicitors' Charity.

Email: rory.okeeffe@rmoklegal.com

Website: www.rmoklegal.com

CHAPTER 25

AI: A DOUBLE-EDGED SWORD

By Izak Oosthuizen
Speaker, Author, Cybersecurity Specialist
London, England, United Kingdom

Some say that likening what is widely considered an emerging technology to a weapon that literally cuts both ways is being a bit over the top. I think not. While the booming AI market offers us amazing benefits such as increased productivity, slick automation, and improved decision-making, it also brings with it a plethora of challenges, including job displacement, ethical and privacy issues, bias, and regulatory non-compliance. Advanced AI models may even have the potential to cause us significant harm through deliberate disinformation campaigns or autonomous lethal weapons, although the savvy ones among us doubt if AI is capable of putting our entire existence in jeopardy. In a Joe Rogan interview, the world's richest earthling admitted that there's "only a 20% chance of annihilation" with AI, but it does pose an existential risk. While I'm far from an existentialist, I am mindful of the fact that artificial intelligence could become an adversary of something close to my heart—cybersecurity—and threaten our digital reality. On the flip side, with guidance and nurturing, AI can be a friendly disruptor and effective security tool, safeguarding the data that we all hold so dear.

As we journey into our technologically driven future, I believe AI will be a high-octane, multi-purpose, automated cybersecurity ally. For those who know me and work with me, to say I like "quick" is an understatement. "Instant" is more on point, and hence my infatuation with automation and anything that reduces human intervention. In cybersecurity, we can leverage AI automation to analyse huge amounts of data in super quick time and detect patterns of activity, suspicious or otherwise. The big plus here is that once an AI model clearly understands what typical behaviour is, it can identify anomalies that may warrant further investigation.

AI automation also means that machine learning (ML) algorithms learn continuously. The more data processed, the better the AI model becomes. Put simply, when generative AI recognises known cyberthreats, malware being a good example, with 560,000 pieces[1] being detected each day, it can help to contextualise threat analysis, making the information more accessible to us by generating explanatory text or images. We can't negate human expertise in cybersecurity, but AI enhances our analytical capabilities, and we can detect threats and take action much more quickly than before. Sridhar Muppidi, CTO of IBM Security, tells us: "Leading organisations are pursuing a forward-looking approach to threat management, adopting AI-powered automation to drive improved insights, productivity, and economies of scale."[2]

Many cyberattacks, including those from malware, ransomware, and credential theft, start with a phishing email that tricks us into clicking on malicious links or downloading infected files. Here too, AI can be a saviour. In the blink of an eye, it can automatically analyse email content and sender details with much greater precision than traditional analytical methods. Another bonus of AI as a cybersecurity tool is in incident response automation, using it to automate security responses such as isolating compromised systems and blocking malicious IP addresses. This means damage to networks is minimised and our data is protected.

I'd like to move on to two specific use cases of AI in cybersecurity and how these benefit our overall security posture. The first of these is one that we typically onboard for many of our clients, endpoint detection and response (EDR) and, by extension, managed detection and response (MDR). MDR offers 24/7 security monitoring, proactive threat hunting, and rapid incident response. It aims to prevent data breaches and ransomware attacks by combining human prowess with the skills of advanced ML models. AI easily beefs up MDR by identifying all the endpoints in use across an organisation. AI systems also ensure that endpoints remain up to date with the latest operating systems and security measures, such as patches and updates.

Specifically, generative AI assistants partner well with detection engineering, the MDR process dedicated to developing, evolving, and tuning detections to defend against evolving cyber threats. From an MDR automation perspective, AI works perfectly for vulnerability prioritisation and managed security information and event management (SIEM). SIEM through advancement in MDR and used in conjunction with AI promotes aggregation, analysis, ingestion, and correlation of logs from different data sources. MDR with AI supports ingestion and deeper understanding of data, unlike any other, to a level where no human can compete with its computational response. It also promotes sentiment analysis and improves reaction time to sub-human levels. With SIEM in play, security teams can proactively respond to cyber threats and stop bad actors in their tracks.

The second use case that I would like us to look at is the synergy between AI and cloud computing. For the laypeople among us, the cloud is basically using the internet to access servers, networks, storage, and databases instead of having them on your physical computer. The popularity of the cloud speaks for itself. Ninety-six percent of companies worldwide use public cloud services, such as Amazon Web Services, and 94% claim that their cybersecurity improved after switching to the cloud.[3] At the moment, 2.3 billion people, more than 25% of the population, use personal clouds such as Dropbox, Google Drive, and iCloud.[4] Cloud computing is big business, and next year that market is forecast to exceed $940 billion.[5]

Since the cloud is undeniably valuable and a lifeline commodity for businesses and individuals alike, keeping it secure is not optional.

Enter AI. AI provides us with visibility into vulnerabilities and risks across multi-cloud environments. This means that security teams can effectively manage and secure our cloud assets. More specifically, AI is pivotal to cloud security posture management (CSPM), where it is used to assess cloud environments, identify misconfigurations, and automatically remediate security issues. AI has also become fundamental to cloud workload protection (CWP), where it monitors and protects cloud workloads from threats, including malware and unauthorised access.

Before we delve into the dark side of AI in cybersecurity, let's briefly home in on more benefits. With cyber threats rising exponentially, growing data sets, and a constantly expanding endpoint attack surface, using AI for detection and response means significantly strengthening your security posture and reducing vulnerability to a cyberattack. It significantly improves the speed and accuracy of detecting critical cyberthreats by filtering through the thousands of logged events in tools like SIEM or MDR to identify those that genuinely pose a risk. AI also simplifies reporting, with generative AI pulling information from multiple data sources to generate clear, concise reports that security professionals can easily decipher and share.

AI helps us discover vulnerabilities such as previously unknown devices on a network, outdated systems, and unprotected sensitive data. By translating complex threat data into natural language, generative AI allows analysts to understand threats more easily and respond effectively, supporting skill development across the security team. Another bonus is that AI provides valuable cyberthreat insights by analysing behaviour across identities, devices, applications, and infrastructure, helping security teams prioritise the most pressing threats and respond with greater precision. Fortinet encapsulates the transformative role of AI in cybersecurity with these words: "AI revolutionises threat detection, automates responses, and strengthens vulnerability management."[6]

I'm also a huge fan of integration. AI has been integrated into numerous cybersecurity tools to enhance their effectiveness. Next-generation firewalls, unlike traditional ones that rely solely on administrator-defined rules, use AI to access and analyse threat intelligence data, enabling them to detect new and evolving cyberthreats. As mentioned earlier, AI-integrated endpoint security solutions help us identify vulnerabilities such as outdated operating systems, detect malware, monitor unusual data transfers, and isolate compromised endpoints to prevent further damage.

The list of benefits goes on and on. AI-driven network intrusion detection and prevention systems can monitor network traffic to uncover and block unauthorised users before significant harm can occur, processing data much more quickly and accurately than traditional systems could. The number of connected internet of things (IoT) devices is estimated to grow to 40 billion by 2030.[7] Many of these are vulnerable to cyberattacks and security breaches, prioritising connectivity over security and using default passwords. AI plays a crucial role in securing IoT devices by identifying threats to individual devices and recognising suspicious activity patterns across enormous IoT ecosystems.

Here's some food for thought. If AI is doing such an awesome job with cybersecurity, what about future employment prospects? We can see that AI will automate many routine cybersecurity tasks, but it will also create new job opportunities in AI security and risk management. To remain relevant, cybersecurity professionals need to focus on upskilling to work alongside AI rather than being replaced by it. The integration of AI-driven analytics will complement, rather than replace, the role of human intuition in cybersecurity. After all, it was humans who gave birth to AI and not vice versa.

AI is great at crunching data and spotting patterns, but being soulless, it can miss context and finer detail and not spot subtle threats like zero-day exploits. We can make judgment calls, algorithms can't. I like to think of AI as a powerful tool in an even more powerful toolkit and a support mechanism for cybersecurity, not a replacement. This statement from Eric Weisburg, a leading data analyst, supports my stance: "Your job security depends less on advances in artificial intelligence and almost entirely on your willingness to make AI part of your personal toolkit."[8]

And now for the other edge of this multi-functional sword. Unchecked, AI has the potential to be a menace to robust cybersecurity. Threat actors are increasingly using the technology to automate, accelerate, and improve cyberattacks. The extent of this sophistication also makes vulnerabilities and attacks harder to detect. Typically, we can think of AI-powered attacks as falling into five categories: malware, particularly ransomware, exploitation of AI vulnerability, phishing, deepfakes, and social engineering. According to the UK's National Cyber Security Centre (NCSC): "AI will almost certainly make cyberattacks against the UK more impactful because threat actors will be able to analyse exfiltrated data faster and more effectively and use it to train AI models."[9]

It doesn't stop there. AI will also enable the rise of low-skilled cybercriminals by giving individuals with little to no experience with coding or hacking tools the ability to craft malicious code with little effort. The NCSC again reports: "AI lowers the barrier for novice cyber criminals, hackers-for-hire and hacktivists to carry out effective access and information gathering operations. This enhanced access will likely contribute to the global ransomware threat over the next two years."[10]

The NCSC is saying that both skilled and unskilled perpetrators can cunningly use AI to identify valuable data sets, execute a ransomware attack, encrypt data, and demand a big payout for decryption keys. For me, seeing ransomware on the rise is an enormous worry, especially in light of the recent devastating attacks on the NHS here at home in the UK, and Change Healthcare and Ascension across the pond. According to Sophos, ransomware payments have increased 500% in 2024, with an average payment of $2 million, up from $400,000 in 2023.[11] After suffering an attack, some companies decide not to pay the ransom, as recommended by the FBI and CIA in the United States, and the Home Office in the UK. Unfortunately, there is much more to a ransomware attack than the immediate financial layout of recovering the data held hostage. Businesses also have to factor in downtime, the cost of operational recovery, and reputation damage.

Another danger is that the bad guys can use AI to exploit inherent vulnerabilities that exist within AI-powered security systems themselves. AI-driven algorithms can scan systems for vulnerabilities, launch automated attacks, and identify and exploit weaknesses, all without human intervention. This makes large-scale attacks more efficient and widespread. Vulnerabilities in large language models (LLMs) and generative AI like ChatGPT can also be targeted through prompt injection attacks, where hackers disguise malicious inputs as legitimate prompts, manipulating AI systems into leaking sensitive data, spreading misinformation, triggering actions through API integrations such as forwarding private documents in an email, and much more. The problem arises here because LLMs find it difficult to distinguish between developer instructions and user inputs. So, using correctly worded prompts, hackers in the know can override developer instructions, and the LLM becomes their servant. A major concern is that, as yet, nobody has found a foolproof way of addressing prompt injection vulnerabilities.[12]

AI has also evolved to facilitate phishing and social engineering. Manipulated AI-generated text and voice technologies can create convincing phishing emails, messages, and phone calls at scale, and we are unable to discern genuine from illegitimate communication or fraud. Attackers can also use AI to include current events, news, and real-time information into phishing emails, making them seem bona fide and believable. AI is capable of generating convincing replicas of legitimate websites, embedding these in emails, and users find it challenging to distinguish between fake and real sites. The sad news is that AI-generated phishing can bypass traditional security measures. Being highly personalised and exploiting our user habits and preferences, they are more likely to succeed. AI has also made social engineering cheaper, faster, and more scalable than ever before. The danger of AI-generated phishing campaigns is succinctly captured in a report on a recent Harvard study: "AI's potential to revolutionise industries comes with significant risks, and phishing is a clear example of this dual-edged sword. AI-driven phishing campaigns are now as effective as human experts."[13]

Thinking back, I remember chuckling at the viral deepfake videos of Barack Obama and Mark Zuckerberg chatting to CBS News about the "truth of Facebook and who really owns the future," and the infamous Nancy Pelosi footage retweeted by President Donald Trump, in which it appeared that the House Speaker was drunkenly stumbling over her words. Actually, it's not that funny. AI-generated deepfakes will make impersonation attacks more dangerous than ever before. Deepfakes aren't limited to impersonation. Troublemakers can use them to spread false information, create misinformation, and confuse us about important political and social issues. Deepfakes can be designed to harass, intimidate, demean, and undermine people. The good news is that we can leverage AI to detect deepfake fraud and verify digital identities with pinpoint accuracy. Businesses and individuals alike need to implement proactive strategies to build resilience against AI-powered deception to mitigate all the risks associated with manipulated digital content.

A further concern is that AI-driven cyberattacks will not just impact businesses, but humanity as a whole through attacks on critical infrastructure such as power grids, healthcare, and finance. Nation-state actors could make use of AI to conduct large-scale disinformation campaigns, economic sabotage, and even digital warfare. The UK parliament highlighted the dangers of how AI could increase the risk of disinformation, saying, "AI can also generate inaccurate text, images, videos and other forms of disinformation, and may lead to increasing risk of online extremism. Foreign state-backed disinformation may aim to provoke confusion, aggravate political polarisation, undermine democracy, or create distrust in societies."[14]

As somebody who many consider to be a cybersecurity aficionado and an IT thought leader, I believe the potential of AI for the security and protection of our data and privacy outweighs the negatives of social engineering, deepfakes, and disinformation. A good start in strengthening your network is to implement zero-trust security models, enhanced by AI for continuous authentication. As we embrace AI, businesses and individuals need to adopt AI-driven cybersecurity strategies to future-proof their security against evolving threats. This is something that I intend to operationalise in my company and with my clients. To my mind, as technology continues to transform our planet, AI will transform how we do cybersecurity. While AI will make cyberattacks easier to carry out and more difficult to detect, we can also harness its power to secure our data and networks, making the world in which we live, work, and play a safer place.

References

1. James Palatty, "30+ Malware Statistics You Need to Know in 2025," Astra, Nivedita, 9 January 2025.
2. Sridhar Muppidi, Lisa Fisher, Gerald Parham, "AI and Automation for Cybersecurity, IBM Institute for Business Value," 3 June 2022.
3. Mariusz Michalowski, "55 Cloud Computing Statistics for 2025," Spacelift, 1 January 2025.
4. Mariusz Michalowski, "55 Cloud Computing Statistics for 2025," Spacelift, 1 January 2025.
5. Mariusz Michalowski, "55 Cloud Computing Statistics for 2025," Spacelift, 1 January 2025.
6. "Artificial Intelligence (AI) in Cybersecurity," Fortinet Inc.
7. "State of IoT Summer 2024," IoT Analytics, 3 September 2024.
8. Eric Weisburg, "You Won't Lose Your Job to AI. You'll Lose Your Job to Someone Using AI, Datos Insights," 30 January 2024.
9. The Near-Term impact of AI on the Cyber Threat," National Cyber Security Centre, 24 January 2024.
10. "The Near-Term impact of AI on the Cyber Threat," National Cyber Security Centre, 24 January 2024.
11. "State of Ransomware 2024," Sophos, 30 April 2025.
12. "What Is a Prompt Injection Attack?", IBM, 26 March 2024.

13. Giorgio Fazio, “AI-Powered Phishing: The Perfect Storm of Persuasion,” Hackernoon, 9 January 2025.
14. “AI, Disinformation and Cyber Security,” UK Parliament, 29 January 2025

About the Author

Izak Oosthuizen is a London IT thought leader, entrepreneur, and cybersecurity expert. He is a member of the Entrepreneur’s Organisation London and was nominated for the prestigious UKTech50 award in 2022. In 2006, Izak founded Zhero, a London-headquartered end-to-end business cybersecurity and IT support company for SMEs. Zhero is a Microsoft Gold partner providing tailored risk mitigation, cybersecurity, cloud, IT support, consultancy, and professional services to a broad spectrum of industry sectors, including medical, finance, legal, insurance, and architecture. Zhero has worked with a diverse range of clients and institutions, including WeWork, Giorgio Armani, Energy UK, Edmond de Rothschild, the Federation of Master Builders, City, University of London, and Dimension Data.

Izak holds several IT qualifications – including various Microsoft, Checkpoint and VMware certifications. Combining his 20-plus years of experience in IT with management qualifications from the London School of Business & Finance and the Cranfield School of Management, he recognises that IT drives growth but also poses associated security risks.

Izak is a respected keynote speaker and has participated in events hosted by The Economist, N-Able, London Market Forums (LMF), and others. Izak is also the co-founder and a director at Cyber London, a not-for-profit supported by the UK Government and the Department for Science, Innovation and Technology. Cyber London’s mission is to make London a centre of excellence for cybersecurity and AI, driving inward investment into the capital.

Izak is the author of the Amazon international bestseller, ‘You Don’t Need a £1 Million Cybersecurity Budget.’ He also co-authored another two best-selling IT books, 'Adapt and Overcome' and 'Cybersecurity NOW', both available on Amazon. These works are essential reads for navigating the ever-changing worlds of cybersecurity, remote working, and the digitised workplace. His insights have also been featured in The Economist, UK Computer Weekly and other prominent publications.

As an individual, Izak is determined and likes to get the job done. His sense of humour makes him an accomplished 'connector', and his humble past allows others to relate to him with a personal touch. Izak's ability to simplify life's complexities instils trust and inspires those around him, both professionally and personally. Izak is an avid proponent of the Biblical philosophy described in Proverbs 11:24: "The world of the generous gets larger and larger…" Izak's inspiration is also derived from Chuck Feeney, the American businessman and philanthropist, known as the "James Bond of Philanthropy." Chuck was 'The Billionaire Who Wanted to Die Broke,' and believed in the joy of giving while living. At the age of 89, he achieved his lifelong goal of giving away all of his amassed fortune – the sum of $8 billion! Inspired by Chuck's generosity, Izak continues to build a company with a Christian ethos and a strong sense of purpose. He aims to bring hope and also play a part in breaking the cycle of poverty, thereby changing future generations.

LinkedIn: https://www.linkedin.com/in/izak-oosthuizen-2509216/

CHAPTER 26

AI FOR SALES AND MARKETING APPLICATIONS

By Elisa Phillips
Sales and Marketing Leader
San Francisco, California

It takes 20 years to build a reputation and five minutes to ruin it. If you think about that, you'll do things differently.

—Warren Buffett

Consistency, transparency, and trust are cornerstones of building and maintaining professional relationships. We have entered a new era where generative AI (Gen AI) technology is reshaping industries and the way we can get things done, at an unprecedented pace. Gen AI is unlocking new levels of efficiency and innovation, and opening up many questions about the future of work, how we connect with one another, and civilization at large. This wave of transformation offers extraordinary opportunities; however it also demands careful stewardship. As we increasingly integrate gen AI into businesses, user experiences, and our daily lives, we must remain conscientious of the short- and long-term impacts of choices made today, which include, and aren't limited to, human and non-human relationships, professional connections, and environmental and natural impacts.

Having been in sales and marketing across my career, I'm excited about the many use cases where AI can be practically applied, offering tangible ways to become more efficient and enhance our relationships and connections. I believe AI can be a powerful assistant if used responsibly.

Practical Ways Sales and Marketing Executives Can Use AI Today

AI is evolving rapidly and there are many practical and accessible ways that sales and marketing executives can apply this technology into workflows right now. It can take on repetitive tasks, surface insights, and free up more time for building authentic connections with customers and partners. As we (the collective "we" of individuals, teams, and companies) invest in AI, the possibilities are endless with how AI could help with workflows, tooling, operations, planning, and so much more.

If you can identify the task you need assistance with or the problem you are aiming to solve, you can leverage AI to help. Think of AI as an extension of your team, a personal assistant and collaborator. Once you get started, AI tools and features become your new default way for activities like streamlining meeting follow-ups, sharpening client proposals, making outreach messaging more scalable or personalized, and much more. To get started, here are a few ways that I use AI, which may help you become more efficient with your existing approaches. Along the way, I'll also share reminders to ensure we use and apply these tools responsibly.

1. Use an AI Companion to Transform Your Video Meetings

In a professional setting, most meetings now happen over video conferencing platforms. AI meeting companions, whether a stand-alone tool or those integrated into the current video conference platform of choice, can transform how you show up and follow-through after your meeting.

You have an upcoming customer meeting with a new client. You've prepared a deck, rehearsed your key talking points, and researched the client's background and latest company news. As the video call begins, you focus intently on building rapport and listening carefully to the client's needs and objectives. Halfway through the conversation, you realize you should be jotting down notes to remember the important context the client is sharing and capture action items discussed—all while staying engaged in the discussion.

Now imagine a different experience: Before the meeting, you activate your AI meeting companion or set it to automatically start at the beginning of your meeting. As the conversation flows, you stay fully present, ask thoughtful questions, listen deeply, and adjust your approach based on the client's reactions without worrying about missing important details. AI transcribes the conversation, identifies key decisions and next steps, and can even flag potential objections based on sentiment analysis. Promptly after the meeting ends, you receive a structured summary highlighting action items, next steps, and key discussion points. This is ready to drop into your CRM and can be used to send a personalized follow-up email and create tasks or reminders for action items. You just got that much more effective at running a client-focused meeting and look polished when you send a follow-up message or have the next discussion.

To get an AI companion, sign up for a new tool (there are many options) or go into the settings of your existing video conferencing platform and turn on the AI companion option. I personally have it set to automatically turn on for all meetings and only have notes be sent to me. Check the settings to ensure it is set up based on your preference.

Stewardship Reminder

Choose AI tools that are transparent about how they store and use your meeting data. In the case you want to record a video meeting, ensure the participants know (in some cases the AI tool will automatically announce this) and ask if they are comfortable with the meeting being recorded. Being transparent is important to maintaining trust.

2. Build Custom Agents to Assist with Writing, Research, and Analysis

One of the biggest developments in AI accessibility is AI agents, which are personalized AI assistants you can tailor for specific tasks. Whether you're drafting a business review, analyzing performance data, or brainstorming creative concepts, a customized AI agent can lighten the load.

Let's say you are preparing for an important follow-up meeting with a prospective client you met last week or developing go-to-market messaging for a new product. You need to get started now so you have time to craft a compelling message that feels personal, aligns with your customer or target audience's pain points, and moves the conversation forward.

You may spend hours digging through meeting notes, researching the latest news, analyzing relevant data, brainstorming how to position your solution just right, and then write and rewrite the email or copy to get the tone exactly right.

When you have an AI agent trained specifically for this kind of task, it can help you speed up this process. I don't believe AI is a replacement for human creativity; however, it's a handy assistant to help you write that email follow-up and do research and analysis on your target customer.

You can create your own agent, for example GPTs from Open AI, or Gems from Google's Gemini, or use agents that are suggested in the platform or that your company may have already created if they are using an enterprise version of an AI platform. When you create or use an AI agent, make sure it's built for a specific task like a "sales meeting prep agent" or a "copy writing agent." The more specific you can make your agent and the more inputs you provide, the more helpful the output will be.

When you're ready to start, open your custom AI agent and upload documents like the client's LinkedIn profile, notes from the last meeting, campaign performance analyses, and bullet points about the solution you're proposing. Provide a clear and detailed prompt to the AI agent with what you are looking to achieve and what you need help with. If you need help crafting a prompt, you can ask AI to help you create a prompt that will give you an optimal output. You may need to try different prompts or go step by step if there are multiple steps you want the AI agent to take. As you and the AI agent learn, it can be a bit of trial and error.

In seconds, your AI agent will draft a tailored email or marketing messaging that considers the context you provided. Review and adjust the output as needed. In a few minutes, you just made yourself that much more efficient.

Stewardship Reminder

AI-generated outputs are a starting point. Review and refine AI output and credit human contributions where appropriate. Additionally resist over-automating messages in ways that feel impersonal or spammy. The goal is to enhance genuine connection, not flood inboxes.

3. Enhance Lead Scoring and Prioritization for Increased Conversion Rates

Sales and marketing teams may have the wonderful problem of having a high volume of inbound leads come in as a result of a marketing promotion or your company's CEO speaking at an event. It can be challenging to identify and focus on the prospects most likely to convert. Traditional lead scoring methods rely on static demographic data which could lead to missed opportunities or time spent on less promising leads. AI can offer a more dynamic and real-time approach to lead scoring and prioritization.

Your marketing team launched a new campaign and many inbound inquiries start coming in. The sales team is eager to follow up; however, they need to decide who to contact first given limited bandwidth and quarterly goals to meet. This is where AI-powered lead scoring can help. AI algorithms within your CRM system, or a standalone AI tool focused on this use case, can analyze an array of data points like website activity, engagement with marketing materials, or firmographics, in real time to predict a lead's propensity to convert. A prediction can provide the sales team with a prioritization starting point. Testing a strategy, gathering data from a cohort of leads, and iterating on the findings will help your team hone the prioritization process. The feedback loop can also be used to refine marketing campaigns for ideal customer profile messaging and offers, and further optimize lead generation efforts in the future.

Stewardship Reminder

Be mindful of potential biases in the data used to train AI lead scoring models. You may need to work with someone in your data team or simply raise the question about the parameters for input data to fully understand the prioritization of the leads.

Looking ahead to the future, the importance of gen AI technologies will only grow as it transforms the way we do things, the way we learn, and the way we interact. Integrating AI tools into sales and marketing processes is necessary, and adopting new habits and processes can provide many advantages. When AI is used well, it can feel like magic.

With any major technological disruption though, we must be mindful of broader implications. Rising and sustained concerns with severe impacts spanning the environment and planet, built-in biases, data privacy, economic prosperity, and much more, cannot be overlooked and require balanced and regulated solutions. Some of the smartest people have been working on AI technologies for decades and stifling innovation isn't in our best interest; however, we are at a tipping point where AI technologies have wider adoption and now is the time to ensure everyone has a seat at the table to determine what the future looks like and what kind of world we're building.

One perspective on ethics and responsibility I go back to often was eloquently communicated by Liv Boeree. Liv is a poker champion and science communicator, and she did a talk called "The Dark Side of Competition in AI" at the first Ted AI conference, which I had the pleasure of attending. Liv talks about the trap we as a society can fall into with AI if incentives aren't aligned for win-win scenarios. The trap is Moloch's trap, which breeds a "mechanism for unhealthy competition." It's up to us, particularly those leaders and government bodies setting the regulations and policies around the use of AI, to build for win-win scenarios that are a "race to the top" (not the top of the market but to the top of outsized benefits for the good of everyone).

Seek out perspectives from those who are focused on building safe technology and advocating for ethical policies and practices. These perspectives shine a light where we need to be focused now as people and companies race ahead.

Ultimately, the AI advantage is about more than just improving business outcomes, it's about using technology in a way that benefits all humans and non-humans, nature and natural ecosystems, and the planet.

About the Author

Elisa Phillips is a dynamic business leader with a career defined by driving digital transformation across media, advertising, and emerging technology platforms. Over her 17-year tenure at Google, she built and scaled early-stage businesses in the US and EMEA. At DoorDash, she fueled hyper-growth of the nascent enterprise advertising business, delivering billions in gross margin value and establishing foundational teams and operations. As COO and vice president of sales at Omneky, a generative AI marketing startup, she doubled revenue and championed customer-driven product initiatives. Elisa joined Uber's advertising organization to support the expansion of its mobility and delivery solutions. She is passionate about innovation, sustainable business growth, and making a lasting impact.

LinkedIn: https://www.linkedin.com/in/elisaphillips/

CHAPTER 27

THE AI-DRIVEN TRANSFORMATION OF THE PHYSICAL WORLD

By Paul Powers
Founder, Physna, 3D AI, Forbes 30 Under 30
Columbus, Ohio

The best way to predict the future is to create it.
—Peter Drucker

It wasn't long ago that artificial intelligence was mostly the stuff of science fiction or at least the kind of technology tucked away in a corner of Silicon Valley research labs. Yet, in a remarkably short time, AI has become an integral part of everyday life—suggesting what we buy online, scanning our job applications, chatting with us through virtual assistants, and even writing articles that appear on our favorite blogs. Now, as AI evolves further, companies such as Physna are demonstrating that the real power of this technology isn't limited to the digital realm; it's changing how we interact with physical objects and supply chains in ways we could only imagine a decade ago.

When most of us think about AI, we picture algorithms crunching numbers, analyzing text, or playing board games at a superhuman level. But the really exciting story right now is how these same technologies are moving from bits to atoms. AI is reshaping physical design and production, setting the stage for a new era—one that's more localized, more efficient, and more personalized.

In this chapter, I'll explore how AI is weaving itself into numerous industries—manufacturing, medicine, construction, supply chain logistics, and beyond. Whether it's helping to streamline assembly lines or develop customized medical implants, AI is empowering us to build, ship, and maintain tangible products with unprecedented precision. Of course, the road ahead will be disruptive, forcing many of us to rethink established business models and question what's possible when automation meets on-demand fabrication.

The Rise of Physical AI

Let's start with the broad shift unfolding right now. We often associate AI with text processing—things like spam filters, chatbots, or predictive text systems. That's no surprise: Digital data is abundant, and it's traditionally been easier to analyze than complex physical objects. That's changing fast, thanks to cheaper sensors, sophisticated 3D modeling software, and a growing web of interconnected systems that funnel real-world data into AI algorithms.

This new era of "physical AI" is about leveraging that data—from precise geometric shapes to minute patterns of wear on a machine part—to optimize how real objects are manufactured or maintained. Picture an airplane component, complete with every rivet, curve, and stress point encoded in a virtual model. AI can simulate the performance of that piece under countless conditions and refine its design before it ever reaches the factory floor.

This is where the idea of a "digital twin" gains prominence. A near-perfect digital model of a physical product provides a testing ground for endless iterations. Once that model is put into an AI-driven system, the design can be honed for cost, weight, and durability—all without risky and expensive real-world trials.

The Bespoke Economy Takes Shape

As manufacturing technology marches forward, one of the most striking trends is the rise of what's often called a "bespoke economy." Instead of pumping out massive quantities of identical products, we can now tailor items to a customer's exact specifications. Traditionally, factories relied on economies of scale to keep costs down, but AI is rewriting that formula in surprising ways.

Imagine a company that makes athletic shoes. Rather than mass-producing a limited number of designs in an overseas plant, then warehousing them, they might use AI-driven production and advanced fabrication (robotic assembly or 3D printing) to deliver shoes built to a customer's unique foot measurements or style preferences. Once the order is placed, an AI system evaluates local material inventories, finalizes the design, and instructs a machine to produce it, drastically cutting down lead times and eliminating wasted stock.

This made-to-order approach doesn't just work for consumer goods. Industrial parts, medical equipment, and other specialized products can all benefit from on-demand customization. By effectively removing the need to retool a factory or retrain staff for every product tweak, AI brings a whole new flexibility to manufacturing. The result is faster response to shifting consumer tastes, minimal surplus inventory, and the possibility of micro-factories located closer to end users.

AI in Medicine and Biotechnology

Healthcare is an industry where even incremental improvements can be hugely significant, so it's no surprise AI is making rapid inroads. Early AI efforts in medicine focused on diagnostics—like scanning images for tumors, something we've proven is far more effective using

3D detection of density abnormalities than relying on images alone—but the technology now touches nearly every corner of healthcare.

Drug Discovery and Precision Therapies

Modern drug discovery often involves screening massive libraries of chemical compounds, a process that historically has been both time-consuming and expensive. AI can simulate molecular interactions, identifying which compounds are most likely to succeed. That streamlines the research pipeline and lets scientists concentrate on the most promising leads. AI also plays a big role in precision medicine, analyzing genetic profiles to recommend tailored treatments for patients, ideally leading to better outcomes and fewer side effects.

Custom Devices and Implants

For prosthetics or implants—from orthopedics to dental—standard sizes dominate. By adding AI-driven design tools to 3D printing, devices can be crafted to match a patient's exact measurements. Surgeons can place a perfectly fitted implant, reducing complications and enhancing recovery times. As hospitals start incorporating on-site AI fabrication labs, they can potentially generate patient-specific solutions on demand, with minimal wait.

Expanding the Manufacturing Playbook

We hear a lot about 3D printing as a flexible, future-oriented technology. While additive manufacturing is key to enabling bespoke designs, AI's impact doesn't stop there. Traditional processes like CNC machining, injection molding, and robotic assembly all stand to benefit from more robust automation and optimization.

Generative Design

Engineers used to iterate designs through a mix of intuition and experience, fine-tuning shapes to balance weight, strength, and cost. Generative design flips that process by having AI propose numerous variations based on specific performance criteria. In other words, the AI does the heavy lifting, producing novel geometries that might be too complex for humans to dream up from scratch. With normalized geometry, we've discovered that AI can actually learn orders of magnitude faster with 3D data than with 2D. This means that the designs can be generated based on a comparatively small number of designs, such as those created by the company in the past. This means generative design can be fine-tuned to the needs and design language of each company.

Predictive Maintenance and Inspections

Another vital but often overlooked area is preventive care for the machines themselves. AI systems that monitor vibrations, temperature, and other performance data can spot early signs of trouble, scheduling maintenance before a minor defect becomes a catastrophic failure. Likewise, automated inspection systems powered by computer vision, and more importantly those powered by a combination of geometric analysis and 3D scanning to improve this, catch flaws on a production line that may be imperceptible to the naked eye. The result is less downtime, better safety, and fewer surprise disruptions.

Collaboration Across Platforms

Manufacturing is rarely a one-company show. It often spans multiple suppliers, production lines, and design platforms. AI tools that can talk to each other—and unify design, production, and logistics data—are invaluable. When everyone from engineers to shipping managers can see the same information in near-real time, coordinating large-scale projects becomes much easier.

Reinventing Construction

Construction might not be the first field you associate with AI, but it's already feeling the effects of data-driven innovation. Anyone who's endured a building project knows timelines and budgets frequently spiral out of control. AI is beginning to chip away at these inefficiencies, transforming how we plan, build, and maintain structures.

Site Planning and Scheduling

Building information modeling (BIM) provides digital representations of construction projects. Layer AI on top of BIM, and you can simulate everything from material scheduling to the ideal sequence for laying foundations. Instead of relying on guesswork, project managers can base decisions on data-driven simulations, shaving off delays and wasted resources.

Robotic Construction

On some job sites, robots are already laying bricks or pouring concrete, guided by AI that ensures tasks are performed accurately and in the right order. Prefabricated components, assembled off-site in automated facilities, can then be transported to the construction location. This speeds up the process and reduces the risk of mistakes on-site.

Smarter, Greener Buildings

AI-assisted design can anticipate how a building will perform long term in terms of energy consumption and occupant comfort. With factors such as wind patterns, local climate, and daily usage data taken into account, architects and contractors can design structures that better regulate temperature or maximize natural light. Over time, such incremental improvements can translate into massive energy savings.

Geometric Analysis Across Industries

One of the key enablers of these AI breakthroughs is the capacity to understand 3D geometry at scale. Solutions from companies like Physna have shown how advanced AI-driven geometric analysis not only helps with product design but also has broader implications. By quickly searching and comparing 3D models—even if they lack consistent naming conventions—teams can locate parts in a shared library that might otherwise be overlooked.

Take an engineer tasked with creating a new bracket for a machine. Instead of designing one from scratch, an AI geometry search can sift through thousands of existing models to find a near-identical bracket. This kind of reuse accelerates product cycles and avoids wasting resources on duplicates. And in an environment where supply chains can be unpredictable, identifying alternate parts that fit the same geometry could be a lifesaver if the original supplier isn't available.

Overhauling Supply Chains

Recent global events have exposed the fragility of traditional supply chains. Natural disasters, health crises, and geopolitical flare-ups can grind production to a halt if key factories or shipping routes are disrupted. AI, thankfully, has multiple tools to make these networks more resilient.

Demand Forecasting

Companies historically relied on past data and educated guesses to gauge demand. AI can blend real-time inputs—economic indicators, social media trends, weather forecasts—to refine those predictions. That means factories produce only what's needed, reducing overstock and the associated warehousing costs.

Dynamic Routing and Logistics

Delays are inevitable in shipping, but AI excels at quickly finding alternative routes or modes of transport. If a port is congested or a trucking corridor is shut down due to a natural disaster, AI algorithms update shipping plans on the fly, preventing costly bottlenecks.

Local Manufacturing Hubs

The logic behind distributing factories closer to end users—rather than concentrating them in low-wage countries—becomes more compelling in an AI-driven world. Small automated plants can produce items on demand, cutting global shipping and reducing the fallout when one plant experiences downtime.

2D–3D Correlation for Part Identification

Here's where Physna has demonstrated a massive impact both for commercial, government, and consumer customers. By correlating 2D and 3D data, they can identify parts that share similar geometry—even if they're mislabeled or have different part numbers. This goes beyond mere visual matching; it trains computer vision models to detect high-precision similarities or variations in 3D objects that aren't fully identical. The implications are huge. Imagine a defense contractor trying to locate a key component that's no longer in production. Or a car manufacturer that needs to find compatible parts after a supply chain disruption. Even consumers stand to gain when repair components can be quickly identified and sourced, preventing extended downtime of essential products. By accelerating these matches and comparisons, businesses and governments can reduce redundancies, avoid delays, and bolster national security through more reliable sourcing.

Environmental and Economic Upsides

One common concern is that automation and AI will lead to massive job displacement. While it's true that these technologies will change the nature of work, history suggests new roles emerge alongside new tools. The challenge is ensuring the workforce can transition—through training and education—to positions like AI oversight, data science, or system maintenance, rather than being sidelined.

On the environmental front, AI may be one of our best hopes for more sustainable manufacturing. When generative design cuts the weight of airplanes or cars, they burn less fuel. When local micro-factories produce goods on demand, huge container ships crossing oceans become less necessary. When predictive maintenance prevents catastrophic failures, machines operate more efficiently for longer. Each small improvement adds up to a more eco-friendly production ecosystem.

Meanwhile, localized, AI-driven facilities can boost local economies and revitalize entire manufacturing regions. Factories that struggled to remain competitive on labor costs might thrive by offering advanced, automated production. Skilled technicians, data experts, and AI engineers become essential hires, breathing new life into areas formerly hit by industrial decline.

Toward Intelligent, Local Production

Imagine a scenario: You drop by a local facility for a custom piece of furniture. You talk through preferences, perhaps using a simple AI interface. The system checks a library of materials and part designs, adapts them to your specs, and sends instructions to an on-site machine. Within a short time, you walk out with the finished product. This vision might have seemed far-fetched 10 or 15 years ago, but it's inching closer to reality.

And it's not just consumer-oriented. Localized production can also work for specialized industrial items, electronics, and even building materials. As AI matures in orchestrating design and manufacturing, it becomes easier to see why producing items near their end destination could become a dominant model. The benefits include reduced transportation costs, faster turnarounds, and a system more resilient to global disruptions.

Of course, reaching that point demands a shift in policy, supply chain partnerships, and workforce training. Companies need clear strategies for integrating AI into every step of production, from planning to distribution. Governments may offer incentives for "greener" manufacturing or revise regulations so that AI-driven plants can operate efficiently without running afoul of outdated codes. Education programs must evolve to equip the next generation with AI fluency alongside skills in robotics, materials science, and design.

The Future in a Nutshell: Charting a New Industrial Path

The pace of change in manufacturing, construction, logistics, and healthcare can feel dizzying, but it also represents a remarkable opportunity to reshape entire sectors in ways that are more efficient, resilient, and consumer-friendly. AI is stepping out of the digital shadows and into the realm of physical production, offering us new tools to tackle age-old challenges—waste, inefficiency, long lead times, and supply chain fragility.

AI-driven 3D geometric analysis is a key, vital part of this transition, showing how to unify, compare, and optimize parts on a grand scale. By quickly identifying geometric relationships that might otherwise be overlooked, these technologies help ensure that manufacturers, governments, and even individual consumers can find exactly what they need, when they need it, without duplication or delay.

Looking ahead, those who embrace AI's potential to reshape physical production stand to benefit from more adaptable business models, lowered environmental footprints, and a workforce trained for higher-skilled, creative roles. Rather than clinging to outdated paradigms, we have a chance to guide this transformation for the common good—just as Peter Drucker reminded us that the best way to predict the future is, indeed, to create it.

About the Author

Paul Powers is a technology entrepreneur and the co-founder of Physna, a company that focuses on 3D search and AI-driven geometric analysis. A Forbes 30 Under 30 winner and AI expert featured regularly in national news interviews and AI keynotes, Paul has dedicated his career to bridging the gap between AI and the physical world. Through companies like Physna and Thangs, Paul's technology career has focused on how AI impacts the physical world, supply chain and manufacturing, and where the digital<>physical realms will soon meet.

Email: ppowers@physna.com

Website: https://www.physna.com/

CHAPTER 28

NAVIGATING THE AI HYPE, DODGING THE CHARLATANS, AND FINDING REAL BUSINESS VALUE

By Gabriele Sanguigno
Founder and CEO of ToothFairyAI
Melbourne, Australia

It is the mark of an educated mind to be able to entertain a thought without accepting it.

—Aristotle

Okay, here we go. You want the real talk on AI for businesses, not the sanitised, venture-capital-fuelled fluff piece you'll find on every second LinkedIn feed. You want it from someone in the trenches, a CEO of a bootstrapped AI company, ToothFairyAI, someone who's seen the good, the bad, and the downright ugly of this AI gold rush. And you want it short, detailed, and without pulling any punches. Well, buckle up, mate, because that's exactly what you're going to get.

This chapter isn't written by some fancy AI model. No, this is me, Gab, tapping this out the old-fashioned way. Why? Because in a world obsessed with speed and productivity gains at any cost, I reckon there's still immense value in slow, deep, messy, and sometimes contradictory human thought. The kind of thinking that leads to real insights, not just algorithmically generated platitudes. It might sound odd coming from the founder of an AI company, but stick with me. By the end of this, you'll understand why a healthy dose of human skepticism is exactly what this AI hype-cycle needs.

So, grab a coffee (Italian style espresso or something stronger, no judgment here), and let's cut through the noise.

Welcome to the AI Circus—Don't Get Eaten by the Clowns

Let's be brutally honest. The current state of AI, especially in the business world, feels less like a technological revolution and more like a three-ring circus on steroids. Every day, there's a new acrobat, a fresh lion tamer promising unprecedented feats, and a whole lot of clowns peddling shiny, worthless trinkets. As the founder and CEO of ToothFairyAI, I've had a front-row seat to this spectacle for years. We've implemented countless AI agents for businesses of all sizes, from construction companies to financial institutions, local councils to recruitment firms. We've seen what works, what's a complete waste of time and money, and, frankly, what's borderline fraudulent.

My team and I have become adept at sniffing out genuine opportunities for AI adoption versus the overblown fantasies peddled by those who stand to gain from the hype. We understand the tech, yes, but more importantly, we understand businesses—real businesses with real problems, real budgets, and real people whose livelihoods are on the line.

So, my promise to you as you read this chapter is this: You're getting the unvarnished truth. This isn't ghostwritten by a PR firm or, ironically, an AI. This is me, drawing on years of hard-won experience, late nights, and more than a few moments of wanting to throw my computer out the window. In a world obsessed with "productivity gains," this is my imperfect human thinking space. The kind that questions, reflects, and sometimes even contradicts itself. It's human, and right now, a bit of genuine human insight is what's sorely lacking in the AI discourse.

Why Should You, a Sane Business Owner, Even Care About AI Amidst This Chaos?

You're probably scrolling through your social media feeds right now, and I'd bet my last dollar you can't go five posts without seeing "AI" plastered everywhere. It's in ads, it's in "thought leadership" pieces (often oxymoronic), it's the buzzword on every consultant's lips. AI has been "the belle of the ball" for the last couple of years, and it's showing no signs of leaving the party.

I've had countless conversations with so-called "thought leaders" (a term I use with increasing skepticism) about the implications of large-scale AI adoption, particularly large language models (LLMs). It's been a wild ride watching opinions swing like a pendulum in a hurricane: from "AI will change *everything*!" to "It's just another dumpster fire like the metaverse," then to a grudging "Okay, it's useful, but it hallucinates and makes stuff up," and now, more recently, a dismissive "Sure, it can read and reason like a PhD student (on surface...), so what?"

Let me be clear: AI is *not* the next crypto fad or a rebranded metaverse. In its current state, even with all its flaws and the surrounding hype, it represents a monumental achievement for humanity. It's already shaping our societies, influencing our consumer preferences, and quietly embedding itself into the fabric of our daily lives, whether we like it or not.

Think about customer service. Do we, as consumers, want 24/7 support from AI agents, or are we happy to wait for business hours to speak to a human? You might think the providers decide this, but trust me, the power lies with us, the consumers. No company in their right mind will adopt AI if they genuinely believe it'll torpedo their revenue or alienate their customer base. They might *talk* a big game, but bottom lines still call the shots.

Consider the Industrial Revolution over 200 years ago. Marginal output gains shifted from being determined by the number of farmers a landowner had, to the amount of energy (coal) and sophistication of machinery available. Power shifted from "Who owns the land?" to "Who owns energy and machines?" We automated human *labour*. AI today? It's about automating human *reasoning and cognition*. That's a profound shift. As computer power becomes ubiquitous on our phones, laptops, and in the cloud, more artificial cognition will be at our fingertips, increasing our reliance on it, whether consciously or not.

So, even if you couldn't give a toss about AI, believe me, AI cares about you. It's being baked into the platforms you use daily to "enhance" user experience, making the AI component a key performance indicator (KPI) of the solution. Social media, search engines, productivity software—it's already happening, and the trend is accelerating.

The AI Gold Rush and Its Many, Many Traps (Seriously, Folks, When NOT to Adopt AI)

Alright, let's get down to the nitty-gritty. The landscape is littered with traps, and I'm sick to death of seeing good businesses stumble into them because they've been sold a pup by someone with a slick PowerPoint and a vested interest. My blood boils when I see responsible, organic AI adoption for businesses derailed by FOMO, driven not by actual business value but by the fear of being left behind in some imaginary race.

The Siren Song of FOMO—Don't Drink the AI Kool-Aid, It's Mostly Sugar Water!

Fear of missing out. FOMO. It's a powerful drug, and the AI industry is mainlining it into businesses worldwide. Not a gentle nudge; it's a full-blown, panic-inducing gold rush, fuelled by an unholy trinity of influencers, VCs, and a media machine thriving on breathless hyperbole. They're all screaming, "Adopt AI NOW, or perish!"

The reality on the ground? While your average punter playing with ChatGPT has skyrocketed, agentic AI systems for business (the kind that actually do useful work) are lagging. Massively. Why? Because it's hard. It requires careful thought, planning, and a clear understanding—not sexy headline material, is it?

This pressure to jump on any bandwagon is dangerous, driven not by a genuine desire to improve, but by primal fear of obsolescence. Here's the kicker: Most of this fear is manufactured. Businesses are better off taking a deep breath, stepping back, and making a measured, strategic decision.

We've seen this movie before. Remember the great cloud adoption stampede? Same damn story. "Migrate to the cloud, or your business will die!" Many businesses, terrified of being dinosaurs, rushed headlong without a plan. Result? A bloodbath. Skyrocketing, unbudgeted costs, complex migrations gone haywire, minimal real return.

And the delicious irony? The same engineers once seen as guardians of clunky, old legacy systems—denigrated for "technical debt"—suddenly became high-demand "cloud architects." Same faces, pricier job titles. Think AI will be different? The "AI engineer" you might now view with suspicion? Rush in blindly, and in a year, they'll be your "chief AI strategist," drawing a bigger salary, managing the same underlying problems with a new layer of AI-branded complexity.

The risks with AI adoption are identical, if not amplified by the sheer hype. Before you even think about implementing AI, conduct a brutally honest assessment: Is AI genuinely the right tool, or are you just trying to keep up with the Joneses or impress board members who've read a Gartner report?

The "Influencer" Plague and the Cult of the Shiny New Thing

Now, let's talk about a particularly infuriating cog in this hype machine: the so-called "AI influencers" and "thought leaders." My god, this is a problem in the industry, a *massive* problem, and it's leading to colossal blunders and wasted effort.

The landscape is crawling with them. Every day, a new guru pops up on LinkedIn or YouTube, breathless with excitement about the *latest* AI model, the *newest* framework, the *most revolutionary* protocol that was, funnily enough, only released 12 hours ago. Their understanding of the technology is often laughably shallow, gleaned from a few press releases and a quick skim of a whitepaper. But they talk with such unshakeable confidence, don't they?

And let's not be naive about their motivations. Many of these individuals have a direct monetary incentive to keep the hype train chugging along. Partnerships, sponsorships, affiliate links—their channels are often plastered with them. They're not offering impartial advice; they're selling you something, whether it's a product, a course, or just their own inflated personal brand.

The real insidious damage here is the "what's new every 24 hours" mentality they cultivate. It creates this constant, frenetic whirlpool of ideas that, crucially, *remain just ideas*. Why? Because the actual implementation of *real* AI agents that do *useful, reliable work* for a business is a grind. It takes planning, analysis of your existing processes and data, rigorous testing, iterative development, and a hell of a lot of hard work. It's not about grabbing the shiniest new open-source model that dropped last Tuesday and expecting miracles. That model that's "slightly better" than the previous one? Guess what—that "previous one" probably came out two weeks prior!

This relentless churn actively prevents anyone—individuals or organisations—from truly sedimenting their knowledge. How can you deeply understand and effectively implement Technology X when, by the time you've figured out its quirks, Influencer Y is already screaming that Technology Z has made X obsolete? It's like trying to build a house on quicksand.

For businesses, this translates into a dangerous cycle of distraction and superficiality. Instead of focusing on solving a core business problem with a well-understood, perhaps even "boring" AI solution, they get caught up chasing the next fleeting trend. They waste time and resources on pilots that go nowhere, tools that don't integrate, and models that aren't a good fit, all because someone with a big social media following said it was "the future."

This isn't innovation; it's a digital form of fast fashion, and it's just as wasteful and unsustainable. We're seeing companies jump from one LLM provider to another, one AI framework to the next, before any real learning or value has been extracted from the previous attempt. The pursuit of the "cutting edge" becomes an end in itself, divorced from any tangible business outcome. And the influencers? They just move on to the next shiny object, leaving a trail of confused and poorer businesses in their wake. It's irresponsible, and frankly, it pisses me off to see so many fall for it.

AI Is a Tool, Not a Magic Wand (And Definitely Not Your Cost-Cutting Engine)

Another pitfall that drives me absolutely bonkers is treating AI as some sort of mystical panacea. It's not. Creating AI agents that are reliable, effective, and actually add value is no different, conceptually, from onboarding and training a good human employee. And guess what? Just like your human staff, these AI agents need to be continuously updated, retrained, and monitored.

You wouldn't hire a new sales rep, give them a ten-minute briefing, and then expect them to triple your revenue by next week, would you? Yet, I see CEOs and business leaders expecting exactly that from AI. They want to plug it in and watch the magic happen. When it inevitably doesn't, they blame the AI, or the vendor, or anything but their own unrealistic expectations and lack of proper groundwork.

An AI agent, particularly one dealing with complex information or customer interactions, needs to be fed high-quality, relevant data. Its performance needs to be meticulously tracked, evaluated, and refined. It's an ongoing process, not a one-time setup. And here's a crucial point: AI agents are not immune to biases—biases present in the data they're trained on or even biases inadvertently introduced by their designers. If these aren't identified and mitigated, they can have serious, reputation-damaging consequences. Imagine an AI loan approval agent that disproportionately rejects applicants from a certain demographic because of historical biases in its training data. Catastrophic.

Businesses *must* cultivate realistic expectations. AI is a powerful tool, yes. It can augment human capabilities in incredible ways. It can automate drudgery. It can uncover insights hidden in vast datasets. But it is *not* a replacement for human judgment, creativity, empathy, or critical thinking, especially in nuanced situations.

Part 2: Finding Real Gold—Successful AI Adoption Done Right (Yes, It IS Possible!)

Alright, enough with the doom and gloom for a moment. Despite the circus, the charlatans, and the monumental piles of BS, AI can deliver incredible value when approached with a clear head, a pragmatic mindset, and a relentless focus on solving actual business problems. It's about finding the real gold amidst all the glitter, drawing from real-world examples. This isn't about chasing fads; it's about methodical, often unsexy, application of technology to make things better.

Consider "The Unsexy but Damn Effective: AI In Customer Service That Actually Helps." Customer service is often the first place businesses think of for AI—it's repetitive and high-volume, a prime candidate. But it's also a minefield if you get it wrong. Get it right, however, and the benefits are substantial.

Let's take the example of an e-learning company we worked with. Their customer service team was, to put it mildly, drowning. They faced a relentless barrage of largely similar inquiries: "How do I reset my password?" "Where can I find this course module?" Their existing chatbot? One of those classic "dumb, scripted, mostly useless" decision-tree nightmares. Deviate slightly, and it just spits out "Sorry, I didn't understand that." Cue customer rage. Business people instantly resonate with that pain. Human agents were spending 80% of their time on these mind-numbingly repetitive queries, leaving little bandwidth for complex issues, proactive support, or value-added interactions. Morale was low, burnout high.

So, what did we do? We didn't just throw another off-the-shelf chatbot at the problem. We architected and implemented a sophisticated customer service AI agent—not just a fancier FAQ. It was built using state-of-the-art (SOTA) open-source large language models (think fine-tuned versions of models like DeepSeek V3 or Llama). Crucially, these models were extensively fine-tuned using supervised fine-tuning (SFT) on thousands of the e-learning company's actual, historical customer service interactions. This was key: The AI learned their language, their common problems, and their specific course details.

Conclusion: Your AI Journey—Pragmatism, Patience, and People First

Alright, so we've dragged ourselves through the hype and pitfalls. What's the final word from a battle-scarred AI CEO? The future isn't humans getting booted by AI agents—that's clickbait. The reality is humans and AI agents working together to build better businesses. As AI matures, smart leaders will integrate AI agents to augment human capabilities.

However, let's not kid ourselves. The breathless pace of AI model improvement is hitting diminishing returns; it's getting expensive for even tech Goliaths to squeeze out tiny gains. This isn't a disaster; it's a reality check. For you, the non-tech leader, this means you can't be a passive bystander waiting for some "miracle model." Get proactive: Leverage your existing data, fine-tune and customise "stock models"—preferably powerful open-source ones—to create AI tools for your specific needs.

This is where open-source AI truly shines. It's your path to retaining ownership, understanding the "intelligence" you're embedding, and avoiding being shackled to the whims of a few hyperscale vendors. It's about democratising AI power, and it's vital.

Sure, self-improving agents are exciting research, aiming to maximise their "reward function." But let's keep our feet on the ground. Defining those rewards for messy, real-world business problems is a monumental challenge. While it deserves attention, don't let it distract from the immediate, tangible opportunities to create value with AI tools available today.

Ultimately, navigating the AI landscape successfully boils down to a few core principles:

1. *Cut through the BS.* Be deeply skeptical of hype. Question everything. Demand evidence of real-world value, not just promises of future glory.
2. *Focus on YOUR problems.* Don›t adopt AI for AI›s sake. Identify genuine business challenges where AI can provide a clear, measurable benefit. Start small, iterate, and learn.
3. *Invest in your data AND your people.* AI is only as good as the data it›s trained on, and its true power is unleashed when it augments skilled, empowered human beings. Don›t neglect either.
4. *Embrace pragmatic partnerships.* You don›t have to do this alone. Work with AI partners who understand your business, who are transparent about capabilities and limitations, and who are focused on delivering real solutions, not just selling software.

The AI journey is a marathon, not a sprint. It requires patience, pragmatism, a willingness to learn from both successes and failures, and an unwavering focus on creating genuine value for your customers and your business. Don't let the circus clowns distract you from the real work at hand. Go build something amazing, thoughtfully.

About the Author

Gabriele Sanguigno is the driving force behind ToothFairyAI, where he serves as CEO and founder, committed to revolutionising global access to artificial intelligence. A mechanical engineer by training with a deep-seated passion for IT and innovation, Gabriele brings over ten years of formidable experience in the AI and startup arenas. He is a leader in generative AI and advanced cloud solutions, with deep expertise in serverless and event-driven architectures. His distinct strength lies in harmonising cutting-edge technical solutions with clear business objectives, leveraging AI and cloud to drive innovation and significant market impact. At ToothFairyAI, Gabriele's mission is clear: to dismantle barriers to AI adoption worldwide, fostering an ecosystem where intelligence is open, transparent, and enriched by diverse human cultures and languages. Originally from Italy, Gabriele embraced Australia as his new home in 2015. When not shaping the future of AI, he enjoys soccer, swimming, and the simple pleasures of great coffee and homemade Italian pizza with his family—his Australian wife, Dian, and their two cherished dogs, Topo Gigio and Tiramisu.

Email: gabriele.sanguigno@toothfairyai.com

Website: www.toothfairyai.com

CHAPTER 29

THRIVING IN THE UNKNOWN: AI AND THE POWER OF ADAPTIVE LEADER AS COACH IN HEALTHCARE

By Linda Sinisi, MS
Founder of Thriving: Strategic Accelerator
Palm Beach, Florida

The ability to learn is the most important quality a leader can have.

—Sheryl Sandberg

The Challenge of the Unknown

The rapid rise of AI presents both immense opportunities and profound uncertainties. We don't know now what we will know in the future about AI, just as past generations could not have predicted the full impact of technological revolutions before they unfolded.

Considering the unexpected between 2020 and 2023, COVID-19 had a profound global impact, leading to significant mortality. In March 2020, the CDC projected 100,000 to 240,000 US deaths—a figure that was quickly surpassed. By January 2023, official reports from Johns Hopkins University attributed approximately 7 million deaths worldwide to the virus, though excess mortality estimates suggest the true toll could range from 19.1 to 36 million, according to a study titled "Coronavirus Pandemic (COVID-19)".

Our World in Data

In the US, COVID-19 was the third leading cause of death in 2020 (350,831 deaths) and 2021 (416,893 deaths), before declining to fourth place in 2022 (186,552 deaths), according to Farida Ahmad, MPH, of the CDC's Division of Vital Statistics at the National Center for Health Statistics. These numbers highlight the pandemic's lasting impact on global health.

The models we rely on to predict the future often fall short, not necessarily because they are flawed, but because they are built upon historical assumptions that may not apply. This phenomenon is heightened when projecting the future of AI. We can estimate its potential, but we cannot fully grasp its long-term consequences. So, how do we navigate this uncertainty to maximize AI's advantages? The answer lies in leadership, adaptive decision-making, and commitment to continuous learning.

Examples of AI in Healthcare

Healthcare is already evolving because of AI. It is helpful to mark where we are. Here are examples of AI in action in healthcare today:

- EveryCure is a non-profit whose mission is to cure rare diseases. They use AI to repurpose drugs. There are thousands of drugs that have successfully gone through clinical trials and have produced the results needed to cure a specific disease. Some of which have been available for decades. Some of these drugs are available at a low price, and unfortunately drug manufactures don't invest in clinical trials unless they see the return on investment. This is where EveryCure repurposes drugs meant to cure another condition. AI scours available medical data on diseases and treatments to uncover potential matches to cure new diseases.

- Health systems are implementing AI-powered scribe solutions, also referred to as "ambient scribe." Today physicians are using these solutions to free them from note-taking while seeing patients. AI captures and transcribes the conversation between patient and doctor and restores the doctor-patient experience to a face-to-face exchange. This AI technology reduces the time that doctors are burdened with documentation after-hours, helping to reduce physician burnout.
- AI-assisted imaging enhances diagnostic precision, streamlines workflows, and improves patient outcomes. At the Mayo Clinic, AI-powered analysis for breast cancer and lung nodules reduces false positives by 5.7% and false negatives by 9.4%. By identifying anomalies more accurately than radiologists alone, the system enables faster, more confident diagnoses, leading to earlier and more precise detection.
- AI-driven real-time monitoring enables early interventions that can save lives. Sepsis, responsible for one in three US hospital deaths, is notoriously difficult to detect early. Johns Hopkins' AI system, Targeted Real-Time Early Warning System (TREWS), analyzes patient data in real time, alerting doctors to potential sepsis cases hours before symptoms become critical. By improving early intervention rates and reducing mortality by 18%, TREWS has been widely adopted and seamlessly integrated into clinical workflows.

- Drug discovery is traditionally slow and costly, often exceeding a decade and billions of dollars. Insilico Medicine leveraged AI to analyze vast biological datasets and identify drug candidates, developing a new compound for idiopathic pulmonary fibrosis (IPF) in under 18 months, a fraction of the usual timeline. This AI-driven approach accelerated clinical trials, reduced costs, and demonstrated AI's potential to revolutionize drug discovery.

Best Practices in Forecasting and AI-Driven Success

Successful businesses don't achieve greatness by having perfect foresight. They excel because they create systems of adaptability—frameworks that allow them to pivot, learn, and evolve in real time.

Businesses that sustain financial success rely on models that account for uncertainty, such as:

- *Rolling forecasts*: continuously updating financial models instead of rigid annual budgets
- *Scenario planning*: preparing for best-case, worst-case, and most-likely outcomes
- *Driver-based modeling*: focusing on key business drivers instead of fixed financial projections
- *Cross-functional leadership*: engaging all levels of an organization to align AI strategy with real-world execution
- *AI and predictive analytics*: using AI not just for automation but for dynamic, real-time decision-making

The Cynefin Framework: A Model for AI in Healthcare and Beyond

The organizations that harness AI effectively are those that embrace uncertainty rather than fear it. They don't wait for all the answers before acting; they create frameworks that allow them to discover new answers as they move forward.

One of the most effective models for decision-making in uncertain environments is the Cynefin framework by Dave Snowden. This model categorizes problems into five domains, helping leaders determine the right approach for AI integration:

- *Clear (Simple)*: best practices apply. Example: AI automating routine administrative tasks in healthcare.
- *Complicated*: requires expert analysis. Example: AI in diagnostics, enhancing radiology and pathology.
- *Complex*: requires experimentation and adaptation. Example: Personalized medicine and AI-driven treatment recommendations.
- *Chaotic*: requires immediate action. Example: AI's role in pandemic response and emergency care.
- *Disorder*: when it's unclear which category applies. The solution? First, define the problem, then apply the right AI strategy.

By applying this framework, leaders can maximize AI's positive impact while minimizing risks, ensuring that AI-driven innovations in the business of healthcare serve their intended purpose.

The Role of Leadership: Creating the Future Together

AI is a tool—but it is subject matter experts who determine how effectively it is used. Here is what great leaders must do in the AI era:

- *Embrace what they don't know*: The most successful leaders don't pretend to have all the answers. They learn by asking the right questions, engaging their teams, and co-creating the future.
- *Develop a culture of experimentation*: Progress comes from iteration. AI success depends on leaders who encourage

calculated risk-taking as a key element of the learning and development process.

- *Act as positive provocateurs*: Leaders should challenge assumptions, encourage curiosity, and foster limitless thinking. The greatest AI-driven innovations will come from those willing to push boundaries.
- *Empower teams at all levels*: The most valuable insights often come from frontline employees, not just executives or external consultants. Leaders must listen, ask skillful questions, and elevate diverse voices.
- *Define and communicate success*: AI is only as effective as the human systems surrounding it. Leaders must help their teams understand what success looks like, providing clarity, acknowledgment, and direction.

When leaders focus on enabling individuals and teams, they turn AI from just a tool into a transformative force.

From Systems to Human-Driven AI

For too long, organizations have built systems that drive people. Now, it's time for people to drive technology.

- AI should unburden people—not to replace them.
- AI should remove barriers—not create new ones.
- AI should be flexible, helpful, and accessible, not a rigid, opaque system that alienates its users.

AI's greatest advantage lies in its ability to augment human intelligence, not replace it. The best AI-driven organizations will be those that elevate human expertise, intuition, and creativity. When we talk about augmenting human intelligence, we mean enhancing human cognitive abilities—such as decision-making, problem-solving, creativity, and efficiency—using technology, particularly artificial intelligence (AI). Rather than replacing human intelligence, AI serves as a complementary tool that expands our capabilities in various ways. Here's what human intelligence augmentation (IA) looks like in practice:

1. *Decision support*: AI analyzes vast amounts of data, identifying patterns humans might miss, helping leaders and professionals make better, faster, and more informed decisions. Example: AI-powered diagnostics assist doctors in detecting diseases more accurately.
2. *Automating routine tasks*: AI handles repetitive, low-value tasks, freeing humans to focus on complex, strategic, and creative work. Example: AI automates data entry, allowing financial analysts to focus on strategy rather than spreadsheets.
3. *Enhanced creativity and innovation*: AI can generate new ideas, designs, and insights that inspire human creativity. Example: AI-assisted tools help designers and writers brainstorm concepts faster.
4. *Real-time adaptation and learning*: AI-powered systems provide personalized recommendations and insights, allowing people to learn and adapt dynamically. Example: AI-driven patient education can adapt to every patient, for example, with learning paced to optimize comprehension.
5. *Cognitive expansion*: AI extends human intelligence by processing complex scenarios and running simulations that would be too difficult for a person alone. Example: AI models predict supply chain disruptions, helping businesses proactively adjust strategies.

At its core, augmenting human intelligence is about leveraging AI as a partner rather than a replacement, making humans more capable, efficient, and innovative in an AI-driven world.

Leader as Coach—Helping Teams Drive Success in a Golden Path Forward

Being a leader as a coach means guiding, developing, and empowering individuals and teams to reach their full potential. Instead of directing or managing, the leader-as-coach model of leadership focuses on fostering growth, building skills, and encouraging autonomy. This leadership style is built on trust, active listening, and asking the right questions rather than the leader driving the conversation and providing all the answers.

A leader-coach is both a mentor and a facilitator, ensuring that their team members not only succeed in their current roles but also grow into future leaders themselves. The leaders who will thrive in this era are those who:

- Listen deeply to understand
- Are authentically curious and ask the right questions
- Challenge outdated assumptions
- Support and elevate the people around them
- Encourage and model collaboration
- Engage in learning together and learning from others
- Support a culture of idea sharing and solution co-creation
- Create workplaces where AI enables human success, not the other way around

Key strategies of leading as a coach include:

6. *Empowerment*: helping individuals understand their strengths and build upon their strengths; sharing optimism and authentic connectedness, whereby individuals, teams, and organizations develop confidence and take ownership of their work
7. *Active listening*: support, appreciation, acceptance, and understanding challenges, motivations, and aspirations

through observation, expressing interest and curiosity, derived through deep listening

8. *Guidance over direction*: encouraging critical thinking and problem-solving rather than dictating solutions
9. *Constructive feedback*: providing insights that help individuals improve rather than just evaluating performance
10. *Understanding*: welcoming thoughts, actions, involvement, and availability; working in partnership based on shared reflectiveness, collaboration, and action-based learning.
11. *Growth-oriented mindset*: encouraging continuous learning, openness to diverse perspectives, thoughtful reflection on challenges, and a focus on sustainable progress rather than short-term gains
12. *Emotional intelligence*: understanding and managing emotions to create a positive, supportive environment. Creating deeper meaningful relationships whereby leaders inspire, motivate, share power and decision-making, and engage teams in creating a shared vision
13. *Leader as role model*: The following are key attributes, leaders exhibit for individuals, teams, and organizations to emulate:

Self-control

- Remaining calm no matter what is going on around you
- Managing your impulsive feelings
- Managing your distressing emotions well
- Staying positive and unflappable
- Remaining focused and thinking clearly under pressure

Trustworthiness

- Doing what you say, when you say you will
- Acting ethically, above reproach
- Building trusting relationships
- Being reliable and authentic
- Admitting your mistakes
- Standing on principle even if it's unpopular

Conscientiousness

- Being thorough, careful, and vigilant; performing tasks well
- Meeting commitments and keeping promises
- Holding yourself accountable for meeting objectives and goals
- Organizing your work

Adaptability

- Changing something or yourself to fit changes around you
- Smoothly handling multiple demands, shifting priorities, and rapid change
- Adapting your responses and tactics to fit fluid situations
- Being flexible, not rigid, regarding how you view things

Achievement-oriented

- Setting challenging self-imposed goals for yourself

- Measuring your performance against your goals
- Actively seeking out information to get the job done
- Using your time efficiently

Initiative

- Taking the lead in problem-solving and conflict resolution
- Taking action to prevent problems in the first place
- Seeking out fresh ideas from a wide variety of sources
- Entertaining original solutions to problems
- Generating new ideas

Final Thought: The Human Advantage in the AI Age

To thrive in the unknown together with the knowledge that we don't yet know the full extent of AI's impact and we can seize the opportunities together entails a recognition of the following:

- We won't see the full potential of AI until we create it.
- We don't know how big the opportunities are—because we have yet to see them.
- We don't know what's possible—because we haven't achieved it yet.

AI is powerful—but human leadership, curiosity, and ingenuity remain unmatched. The real AI advantage is not just in the technology itself. It's how we choose to use it. Will we use AI to deepen our understanding, enhance human potential, and create unprecedented opportunities? Or will we allow it to limit us, replacing our own decision-making, creativity, and leadership? AI is not the future—we are. And the future belongs to those who are willing to step into the unknown, embrace uncertainty, and shape what comes next. The choice is ours; the unknown is not a barrier—it is an open field of possibility. And the future will belong to those who dare to step forward and build it.

About the Author

Linda Sinisi has over four decades of success in business and technology leadership. A former CIO within the Veterans Administration—Philadelphia and Penn Medicine-Pennsylvania Hospital with senior director roles in Siemens Health Services, Linda built strategy, business transformation, and governance and change management skills during her accelerated career path. Her success is grounded in her unrelenting support of her clients' goals, both organizational and personal, and her uncanny ability to unlock their potential to achieve superior results.

Since 2010, Linda has been applying her extensive, firsthand executive leadership experience in support of her clients' leadership development needs, offering coaching, workshops and consulting services. She is the Founder and Chief Executive Officer at **THRIVING,** ***an organization dedicated to supporting leaders and organizations as they:***

- Expand their leaders' capabilities
- Improve influencing, communication and collaboration skills
- Utilize tools to unlock and harness the full potential of their employees

Linda's unique and enthusiastic approach to professional development enhances business outcomes by motivating and empowering her clients to immediately apply their learning.

Linda is a Professional Certified Coach and also holds a Certification in Leading Change, from The University of Pennsylvania with a Master's Degree in Organizational Dynamics.

Email: Linda@ThrivingPB.com

Website: www.ThrivingPB.com

CHAPTER 30

THE IMPORTANCE OF TASTE IN THE AGE OF AI

By Greg Starling
Head of Emerging Technology, Gen AI
Oklahoma City, Oklahoma

Technology alone is not enough—it's technology married with liberal arts, married with the humanities, that yields us the results that make our heart sing.

—Steve Jobs

I recently got into an argument with a comedian. Trust me, if you find yourself in a similar situation, it's not worth it. I was speaking to a group of content creators about using AI to offload the more labor-intensive and costly aspects of their creative processes. The crux of the conversation was how to use AI to amplify their voices, how the barrier to entry has become democratized. It was intended to be an uplifting talk about not needing a massive budget to bring what is in your head into the world.

Generally, the group was receptive to the conversation, but the comedian, not so much. He used his talent of turning a phrase to try and turn the room. His resistance was understandable. He was experiencing the same existential crisis many of us experience the first time we encounter AI—where do I fit into this new world? Am I going to lose my job? Is there a place for my creativity?

The Heart of Creativity

That's a big question. In this new reality where AI is encroaching on things that not long ago were the sole domain of humans, is there still a place for our creativity? After all, nothing is more core to the human experience than our desire to be creative. But rest assured, AI is still far away, maybe a forever away, from taking that from us.

AI is a craftsman, skilled at observing and mimicking what creators do. But AI is not a creator, at least not in the way humans are. We bring something unique to the creative process: taste. That sense of knowing when a piece is complete, when to add a final brush stroke, or when to delete that last sentence. It's this quality, born from our lived experiences, emotions, and struggles, that AI cannot replicate. This taste isn't innate—it's forged through years of trial and error, through the wisdom of failed attempts and unexpected successes, through the deep understanding that comes from walking the path of creation ourselves.

Understanding AI's Role

This quality of taste, knowing how and why to make creative choices, is at the heart of what separates human creativity from artificial intelligence. When people ask what AI can do, I'm more interested in turning that question on its head and asking, "What do you want it to do?" That's the key difference. AI doesn't want anything. It has no desires, dreams, or drive. It has patterns, probabilities, and processing power.

Think about the last thing you created: a presentation for work, a few thoughtful notes in a birthday card, or even just a great social media post. What drove you to make it? There was probably a spark, a moment where you thought, "Wouldn't it be cool if ..." or "This needs to exist." That spark is what AI can't replicate. It can help you build the fire, but it can't strike the match.

This is where taste comes in. Taste isn't just knowing good from bad; it's about having a vision and knowing how to get there. It's about making purposeful choices. When a chef adds a pinch more salt, when a writer cuts their favorite paragraph, when a designer chooses one font over another, these are more than technical decisions. They're choices guided by an internal compass pointing toward something specific.

AI can analyze millions of recipes and tell you that salt enhances flavor. It can study countless books and tell you that shorter paragraphs improve readability. It can process every font ever created and tell you which ones are used in luxury branding. But it can't tell you what your restaurant needs to stand out in your neighborhood. It can't know which paragraph, though beautifully written, distracts from your story's emotional core. It can't understand why that slightly quirky font perfectly captures your brand's personality.

Because here's the truth about creativity: It's not about making something perfect. It's about making something meaningful. Every creative choice serves a specific vision, and you can't serve a vision you don't have.

The Evolution of Creative Tools

The fear of obsolescence isn't new. The comedian's reaction echoes through history, from painters who feared photography would make their craft irrelevant, to musicians who worried recording technology would eliminate live performances. In 1840, after seeing his first photograph, French painter Paul Delaroche declared, "From today, painting is dead." A widely circulated magazine of the time, *Brush and Pencil*, lamented, "Photography would in time entirely supersede the art of painting." The real angst of that time is mostly lost, but those fears were very real. Yet these technologies didn't replace art; they became new tools for making it.

But AI feels different, more personal. When a machine can generate a decent poem or sketch a portrait in seconds, it strikes at something deeper than our professional pride. It challenges our core belief that creativity makes us special. That's why the existential crisis hits hard. It's not just about losing our jobs; it's about losing our place in the story of human expression.

The Timeless Drive to Create

Cave paintings weren't just art, but declarations of existence. They declared, "I was here. I saw this. I felt this. I mattered." That's the real fear AI stirs in us—if a machine can make something that looks like art, how can I leave my mark? But these ancient works reveal something profound about human creativity that remains true today: We create not because we're the only ones who can, but because we must.

Those early artists didn't wait for perfect tools. They used what they had to express their lived experience, their fears, their triumphs, their understanding of the world. They created because making things helps us process being alive, make sense of our existence, and share that understanding. The tools evolved from charcoal and cave walls to styluses and screens, but the underlying drive remains unchanged. Each technological advance, from the invention of paint to cameras to AI, has simply given us new ways to pursue that fundamental human need to express, connect, and create meaning.

AI isn't here to replace that drive. It's just the newest tool in humanity's creative toolbox, offering new possibilities for expression while leaving the essential human elements—vision, taste, and the need to create—unchanged. Just as photography freed painters to explore new forms of expression, AI tools can liberate creators to focus on the uniquely human aspects of their craft.

A Farm and a Field

I have a farm, or at least what a city kid thinks of as a farm—31 acres of pastures and groves. Yes, a real farmer might call it a "farmette" or a fun little hobby, but for us, it's an enormous amount of land. Up until we bought it, cattle had roamed the acreage for over 100 years, plodding through the fields in the rain, leaving hoof-shaped depressions across the pasture. That makes for an incredibly bumpy mow. And a bumpy mow means a slow mow: twelve hours to get the grass knocked down kind of slow.

We should dump some dirt and get a box blade out there to smooth the field. It would cut mowing time by down 25% to 30%. AI can provide similar gains. AI is your dirt and box blade, smoothing out the bumps of content creation.

AI aids research, structures content, writes drafts, and provides feedback. When used effectively, it amplifies your creative vision and refines your ideas. AI allows you to focus on the core creative task, taste. But AI doesn't know when to stop. It doesn't know when enough is enough. It will endlessly smooth the field. You have to know when it's smooth enough, when one more paint stroke or an extra sentence turns something elegant into something distasteful. That ability to recognize the perfect moment of completion, that's the human element AI can't replicate.

The Path Forward: Approaching AI with Taste

Consider how photographers work with RAW files. The unprocessed image isn't the final product; it's the starting point. The artist's vision emerges in the editing: which details to enhance, what mood to create, how to guide the viewer's eye. That's how we should approach AI.

When writing, AI can organize research, suggest angles, and generate rough drafts. But that's all they are, rough drafts. The real work begins when you shape that material to match your vision. You might keep 10% of what AI generates or use it to test approaches before finding the right one. The key is maintaining control of your creative direction.

AI isn't about replacing creativity but eliminating friction. When researching a topic, AI can quickly summarize key points from multiple sources. When stuck on structure, it can suggest ways to organize thoughts. These tasks used to consume hours of creative time. Now creators can focus on crafting the message, fine-tuning the tone, and ensuring every word serves a purpose.

Shaping the Future of Creative Expression

The relationship between human creativity and AI tools will continue to evolve. Just as today's artists blend traditional techniques with digital tools, tomorrow's creators will find new ways to incorporate AI into their processes. This evolution isn't about replacing human creativity but expanding our creative possibilities.

The key is understanding that creativity isn't just about the final product. It's about the journey of discovery, finding your voice, and translating your unique perspective into something that resonates with others. AI can assist, but it can't replace your experiences, your emotions, or your understanding of what moves people. These are the irreplaceable elements that give your creations their power.

AI is an efficient, tireless research assistant. They'll gather the materials you need, but they won't write your story. They can't, because they don't know your vision. Only you do.

The future of creativity isn't about AI replacing human vision. It's about smoothing the path to it. Remember that bumpy farm field? Once it's leveled, you're not just mowing faster; you're mowing better. You can focus on the perfect cut height, following the land's contours, and creating satisfying straight lines that turn your field into a park.

That's what AI offers creative people: a frictionless experience that allows our creativity to flourish. You can focus on refining your voice, perfecting your message, and adding subtle touches that turn good work into great work.

Because here's what the comedian missed: AI isn't here to tell your jokes, write your songs, or paint your paintings. It's here to help you tell more jokes, write more songs, and paint more paintings. The human drive to create, that spark that made us draw on cave walls and tell stories around fires, isn't going anywhere. We're just finding better tools to leave our mark.

No matter how smooth technology makes the field, only you can tell your story. And in an age where AI can generate endless content, it's your unique vision, your taste, and your human experience that will make your creations stand out and resonate with others. The future belongs not to those who use AI most effectively, but to those who combine its capabilities with their own irreplaceable creativity.

About the Author

Greg Starling is the head of emerging technology at Doyon Technology Group, where he explores the frontier of innovation. A recognized thought leader for over two decades, Greg has shaped conversations around technology, leadership, and management. His work has been featured in *Forbes, Wired, Inc., Mashable,* and *Entrepreneur,* among others. As an Inc. 500 entrepreneur and holder of multiple patents, he has twice been named Innovator of the Year by *The Journal Record.* Greg also leads one of the world's largest AI information communities, helping professionals navigate the rapidly evolving landscape of artificial intelligence.

Email: gstarling@gmail.com

Website: www.gregstarling.com

CHAPTER 31

GOING FROM CODER TO COMMANDER

By Jodessiah Sumpter, MBA
Innovation-Focused Tech Executive & Creator
Alpharetta, Georgia

Do you see a man skillful in his work? He will stand before kings; he will not stand before obscure men.

—Proverbs 22:29

The Technical Ceiling

Elevating yourself from the role of software engineer or developer performing daily coding tasks to a technology executive or thought leader is challenging. While you will see leadership positions seeking candidates with hands-on coding and production deployment experience at scale, the reality is that the skills that will help you excel as a developer are different from those required to be a highly effective executive. Software developers, also known as coders, are often shown in movies as the geeky person sitting in their cubicle in the corner barely able to hold a relevant conversation with their friends. In more recent movies we are seeing the attractive "cool" techie able to hack into the systems across the world with ease while being revered by the people seeking their help. Despite this new view on the value of developers in cinema, the personality traits of introversion, deep thinking, and focus on technology deliverance versus business impact is closer to reality. As a software developer, your superpower of focused isolation for coding means you are limited to a career ceiling of staff engineer. Then from here likely begins the endless journey of salary hopping to different organizations just to keep up with inflation.

The road to becoming an executive at an organization is often not smooth or straight for someone with deep technical skills. While there is strong evidence of success at large tech companies for former hands-on tech leaders like Larry Page at Google and Bill Gates at Microsoft, this is viewed as the exception and not a strict requirement for executive leadership. Developers are primarily focused on being logical problem solvers and system builders of technology. Executives are expected to be vision-defining, people-inspiring strategists who can drive diverse teams to extraordinary results. As developers advance in their careers and begin to seek leadership roles, these differences in expectations become a major roadblock to their success.

The advent of artificial intelligence (AI) has provided a pathway for developers to navigate their way to leadership roles. In this chapter we will review the potential impact of AI on the career of a developer and how it can help alleviate the common struggle of technical software engineers hitting a "leadership ceiling." You will learn how AI could serve as a bridge to the next level for developers who leverage it correctly. Let's start with the developer mindset.

The Mindset Shift—From Builder to Visionary

The transition from focusing on building a solution to envisioning the future in a clear, communicative manner can be challenging. As a coder you are encouraged not to "scope creep" and to limit your futuristic thought to making accommodations for a scalable architecture.

Most of the time you would not have a deep understanding of the business as a whole but rather be engaged in the specific area that the system you're developing impacts. When a developer transitions to an executive, they move from the thought of "How do I build this?" to "Why should we build this?" This is a major shift to the mindset of a developer who is judged on getting the solution to work versus caring why the solution is even necessary to the business.

Making this transition from doer to communicator will push the average developer to their breaking point. The use of soft skills like effective storytelling, the power of influence, and strategic collaboration contrast with the personality traits of a typical developer. They will need to step out of their comfort zone and embrace training for the skills necessary for executives. While things like mentorship, executive training courses, leadership books, and advanced degrees can help developers make the transition, AI can offer another tool in their toolbelt to support their ascent to leadership.

AI as a Communication Co-Pilot

There are several artificial intelligence tools that can be leveraged for developers who are aspiring to be executives. First let's look at some the key skills that a developer would need to master to increase their odds of making the leap to executive:

- *Executive summaries and reports*: A developer would need to provide well-written articles and reports for management, their peers, and C-suite leadership.
- *Email drafting and meeting preparation*: A developer would need to learn effective copywriting and meeting management skills.
- *Strategic analysis and research*: A developer would need to be able conduct research and analysis of the industry and business that they are in.

- *Public speaking and storytelling*: A developer would need to be able to present a clear vision and communicate a story of their views of the future.
- *Partnerships*: A developer would need to identify and establish collaborations to drive their vision forward.

AI Tools like ChatGPT provide developers an opportunity to tailor their questions through effective prompt engineering to help them communicate better. For example, ChatGPT can write messages, draft emails, and conduct research on topics of interest. If you want to get extreme, you could train your voice on ElevenLabs.io, so an AI can answer your calls or speak on your behalf in virtual meetings. Effectively using AI to review or guide communication will help a developer become better and more confident in taking on leadership tasks.

Visualizing Complexity—Making Architecture Talk

The importance of the use of visuals in storytelling through effective diagram and image generation cannot be understated. Software developers often struggle with the visual presentation of their logical (i.e., left brain) thoughts and solutions since it requires a more creative (i.e., right brain) mindset. Now with highly creative AI image and video generation tools, these developers can leverage their strength of logical problem solving through detailed prompt engineering of the creation request. Their detailed and specific nature becomes an advantage since many AI image generation platforms perform better with specific, clear, sequenced guidance towards the object that is to be created. In addition, the same prompt can often be used repetitively, yielding different unique results.

While the default players for AI generation like ChatGPT, Perplexity, and Claude do a decent job now for visual creation, it is often better to utilize tools that specialize in the areas of communication needed for the best results. The visuals developed are easier to define in these tools while offering additional benefits like collaboration ability and enhanced layout alternatives. Here are some recommended AI tools that can be used to elevate a developer's visual communication effectiveness:

Diagram generation: AI tools for whiteboarding or to create system diagrams from natural language:

- Mermaid: https://mermaid.live
- D2: https://play.d2lang.com
- BoardMix : https://boardmix.com/ai-whiteboard/

Pitch decks and presentations: AI tools to assist in turning technical architecture into visually digestible slides for executive audiences:

- Slidebean: https://slidebean.com
- Gamma: https://gamma.app
- SlidesGo AI: https://slidesgo.com/ai/presentation-maker
- Beautiful AI : https://www.beautiful.ai

Collaborative ideation: AI tools that aid in whiteboarding sessions and brainstorming future-state architectures with non-technical stakeholders:

- Ideamap: https://ideamap.ai
- MIRO: https://miro.com/ai/
- Canva Whiteboard: https://www.canva.com/online-whiteboard/

Now that we have the AI tools to help tell a compelling story, let's explore ways to ensure the right business insight and strategies are embedded in the communication.

From Data to Insight—AI for Business Thinking

Many of us have heard the phrase or were told as kids that "Knowledge is power." This is especially true in the US where societal movements around education reform, social injustice, and thought enlightenment have guided people towards the acquisition of more and more education. We are pushed to believe that being right is what matters most, and we should work hard to find the correct answer. Software developers are wired to drive towards seeing the problem and finding the "right" answer to the solution. This thought process leaks into their interactions with others where they need to prove that they are right. What developers aspiring to make the leap to executive sometimes fail to realize is that great leadership isn't about being right; it's about getting it right at the right time.

Effective use of data for business insights on what is important for the business today versus the future is a skill that corporate executives previously acquired through mentorship, top university education, social connections, or early-career trial and error. Deep knowledge of the core areas of the business required more soft skills since it centered around more of a knowledge transfer through direct communication with people who had experienced success in their field of expertise. You had to pay for access to these experienced people and their coaching or have the right connections to hopefully get the privilege of time with them.

With the introduction of AI, this limitation is on the pathway to elimination. As a developer you can spend time asking for the questions that you need to be able to answer. Here are a few areas of focus that a developer can explore with AI:

- *Scenario analysis and forecasting*: using AI for "what-if" business modeling and customer impact projections
- *Product-market fit and customer empathy*: how AI can simulate customer personas or feedback to help developers think beyond code
- *Competitor benchmarking*: leveraging AI to analyze markets, competitors, and trends to position technical solutions strategically

Now that we have the mindset, the communication methodology, the visuals, and the data, let's explore practical ways we can adopt AI effectively into the life of a developer.

Practical Adoption Pathways

Now that we have walked through the power of leveraging AI in your leadership development journey, let's discuss how to add it to your toolkit of success. You should identify the AI tools you want to explore from the previous lists provided and outline a plan for exploration. I recommend starting with ChatGPT first to analyze your specific next steps to leadership. Here are some suggested prompts to help you get started:

1. Define the next steps in your career based on your current status:

   Prompt example: You are a brutally honest career consultant from Korn Ferry International.

   My goal is to grow a career to become CTO at a Fortune 500 company and my income to be at least $400k by December 2026, but I feel stuck.

   Here's what I enjoy doing:

   - Building amazing emerging technology
   - Attending tech conferences
   - Scaling new ideas from MVP to production
   - Working from home at least two days a week

   You have my resume, so please leverage it to see my roles and experience.

   The monthly income of my latest role is $20,000.

   I have a family of five and serve as the primary breadwinner.

   Act like a Korn Ferry advisor.

   - What's blocking my next level of growth in my career?
   - What do I need to STOP doing immediately?
   - What do I need to START doing now to unlock a higher income?

- Where am I playing small without realizing it?
- What three moves should I make in the next 30 days to hit my goal?

2. Identify opportunities for improvement with a detailed plan to get better

   - *Prompt example*: I am working toward the goal of becoming CTO of a Fortune 500 company in the next five to ten years. Based on my resume and information you have about my skills, can you outline a detailed plan on how I can accomplish this goal? Please include events, conferences, executive coaches, groups, etc., that I can leverage to increase the speed of me meeting the goal. Also please outline a free versus paid path to move faster with estimated costs and results.

3. Go big by adding specific financial and/or life goals.

   - *Prompt example*: Can you tell a story of my future with me becoming a millionaire in the next X years within my current industry? Leverage my experience and expertise to outline a specific executable plan starting today in order to make it happen. Please include an outline of executable steps for me to take to make it happen and include daily affirmations I can use to stay encouraged and focused.

Security and Ethical Usage of AI

The AI prompt examples above are focused on you specifically. They provide detailed steps based on saved knowledge from the details you provided. This makes it extremely tempting to just load company data into AI systems to get the most accurate and tailored results. Please DO NOT DO THIS! Your company cyber security, HR, and/or data security team should have provided you detailed instructions on how to handle sensitive and confidential data within your organization and which AI tools are approved for ingestion of that data. Many companies run their own hosted internal versions of approved AI tools and communicate what level of data can be inputted into them.

What approach should you take in order to maximize the usage of AI tools for transition from coder to commander? Here are a couple of suggestions: First, ensure the AI tool you want to utilize is an approved solution within your organization. Verify with your management if it can be used and with what level of data (i.e., public, sensitive, confidential, etc.). Second, do your best to not include actual production or confidential information in the AI tool. This requires some thought, but there are ways to phrase questions or leverage fake data to get the response that you need.

Elevate to Commander by Leaning into Your Coder Strengths

The use of AI is not about transitioning a developer into a non-technical person. It is about elevating their skillset and amplifying their ability to lead. With the addition of AI a developer now has intelligent assistants that can be called upon whenever there is a need. As long as AI is utilized in an ethical, secure manner, a developer should view it not as a threat to their identity, but as an assistant that elevates their impact. Define a clear vision, then choose your AI tools wisely, and you will be on your way from coder to commander.

About the Author

Jodessiah Sumpter is an award-winning innovation and technology executive with over 25 years of experience leading emerging technology initiatives across startups, Fortune 500 companies, and mission-driven organizations. As the founder of Tech Levitate, Joe helps businesses harness the power of AI, gaming, and digital transformation to scale faster and build what's next. He has built and led high-impact teams in software engineering, product strategy, and innovation design, including formerly serving as director of emerging technologies at Starbucks. In addition, he collaborates with global technology leaders such as Microsoft, NVIDIA, and Google to conceptualize and deploy cutting-edge technology concepts.

Previously, Jodessiah served as the innovation lead at NCR Corporation, where he spearheaded the development of over 20 patents and delivered innovative solutions across hospitality, retail, and smart branch pilots. With a career spanning senior roles in technology at organizations like Home Depot and In Touch Ministries, he has consistently demonstrated expertise in emerging technologies, software engineering, and technical leadership.

Jodessiah is a published author of the book *Unity in Action: Multiplatform Game Development in C#*. He is leveraging this knowledge to create games that blend emerging technologies such as AI and interactivity to create next-gen immersive digital experiences. Jodessiah has obtained a Bachelor of Arts in Economics and Master of Business Administration with a concentration in managed information systems from the State University of New York at Buffalo, plus earned a post master's certificate in marketing from the University of Dayton.

Email: joe@techlevitate.com

Website: www.techlevitate.com

CHAPTER 32

ETHICS IN AI: WILL WE LIVE IN A UTOPIAN OR DYSTOPIAN FUTURE?

By Sakina Syed, B.Sc.
AI Consultant and AI Engineer
Toronto, Ontario, Canada

AI will be the best or worst thing ever for humanity.

—Sam Altman, CEO of OpenAI

The Fork in the Code

Artificial intelligence is advancing at an unprecedented pace—transforming industries, reshaping economies, and redefining how we live, work, and connect. From voice assistants to autonomous vehicles, AI is no longer a distant vision of the future; it is a powerful force embedded in our present.

But with this power comes a profound ethical dilemma: Will AI lead us to a better world—or a darker one? Will it be a tool for justice, innovation, and human flourishing, or a mechanism of control, inequality, and harm?

This chapter explores the two divergent paths that lie ahead. On one side is a utopian future, where AI is developed and deployed ethically, enhancing lives and expanding opportunity. On the other is a dystopian future, where unchecked AI exacerbates societal divides and undermines human dignity.

The outcome is not predetermined. It depends on the choices we make now—how we design, regulate, and relate to the intelligent systems we are creating. The question is not just what AI will become, but what kind of future we are willing to build.

The Ethical Foundations of AI

Core Ethical Principles

At the core of ethical AI development is a set of guiding principles that help ensure technology is designed and used in ways that genuinely benefit society. These principles—beneficence, non-maleficence, autonomy, justice, and transparency—are not just abstract ideals. They are practical tools that help developers, policymakers, and organizations evaluate whether an AI system is acting in the best interest of people.

Beneficence means that AI should actively promote well-being and contribute positively to human lives. For instance, an AI system used in healthcare should aim to improve patient outcomes, enhance diagnostic accuracy, or streamline patient care delivery. But doing good is only one side of the coin. We must also ensure that patients have their privacy respected and have their data secured.

Non-maleficence, or the commitment to "do no harm," reminds us that AI must also avoid causing unintended damage. This includes preventing biased decision-making, protecting user privacy, and ensuring that systems don't reinforce harmful stereotypes or inequalities.

Another important concept, autonomy, emphasizes the importance of respecting individuals' rights to make informed decisions about how AI affects them. People should have control over their data and be able to understand and challenge decisions made by automated systems. While this sounds like it should be a given, in practice, many AI systems operate as black boxes—leaving individuals with little insight into how decisions are made or how to contest them. Bridging this gap is essential to uphold autonomy and foster trust in AI technologies.

To support individual autonomy in AI, several key strategies must be implemented. First, users should have clear control over their personal data through informed consent and the ability to access, download, or transfer their information across platforms. Second, AI systems must be transparent and explainable, offering understandable reasons for decisions and using interfaces that help users grasp how the technology works. Third, individuals should have the right to challenge AI decisions, especially in critical areas like healthcare or employment, with accessible channels for appeals and oversight. Finally, ethical design practices and strong legal frameworks—such as the GDPR and emerging AI regulations—are essential to ensure fairness, accountability, and respect for human dignity.

Justice calls for fairness and equity in how AI is developed and deployed. This means ensuring that benefits and risks are distributed fairly across different groups and that no one is unfairly disadvantaged by the use of AI technologies.

Finally, transparency is about making AI systems understandable and accountable. Users and stakeholders should be able to see how decisions are made, what data is used, and who is responsible when things go wrong.

When these principles are thoughtfully embedded into every stage of AI development, from initial design to real-world deployment, they help ensure that technology serves humanity rather than undermines it. They provide a moral compass for navigating the complex and rapidly evolving landscape of artificial intelligence.

In a utopian world, where these values are universally upheld, AI becomes a trusted partner in building a more just, creative, and compassionate society, amplifying our best qualities rather than replacing them.

The Role of Human Values

AI does not exist in a vacuum—it reflects the values of the people and societies that create it. Cultural, social, and philosophical perspectives deeply influence how ethical concerns are interpreted and prioritized. For instance, Western frameworks may emphasize individual autonomy and rights, while Eastern philosophies might focus more on collective harmony and social responsibility. These differences shape everything from data privacy norms to acceptable uses of surveillance technology.

A clear example of this is the contrast between the European Union's General Data Protection Regulation (GDPR), which strongly protects individual data rights, and China's Social Credit System, which integrates AI and surveillance to promote collective behavioral norms. Both systems are driven by ethical reasoning, but they reflect very different cultural priorities. Recognizing and respecting this diversity is essential for building AI systems that are not only technically robust but also ethically inclusive. A truly ethical AI must be adaptable to different cultural contexts while upholding universal human dignity.

The Challenge of Alignment

One of the most pressing ethical challenges in AI is the problem of alignment—ensuring that AI systems act in ways that are consistent with human values and societal goals. As AI becomes more advanced, autonomous and complex, the risk increases that it may pursue objectives that diverge from human intentions, even if unintentionally. This misalignment can manifest in subtle ways, such as reinforcing social biases, or in more extreme scenarios, like the misuse of AI in autonomous weapons or also in algorithms used for decision-making.

In another example, a hiring algorithm trained on historical data might learn to favor male candidates over equally qualified female applicants, not because of explicit programming, but because it mirrors past biased hiring practices—thus perpetuating inequality. Solving the alignment problem requires interdisciplinary collaboration, robust testing, and continuous oversight. It also demands humility: We must acknowledge the limits of our foresight and build systems that can adapt to evolving ethical standards and societal needs.

The Utopian Future: AI for the Common Good

AI as a Force for Equity and Empowerment

Artificial intelligence holds the transformative power to become one of the greatest equalizers of our time—narrowing deep-rooted disparities in education, healthcare, and economic opportunity.

In classrooms around the world, AI can tailor lessons to each student's unique pace, learning style, and language, turning a smartphone into a personal tutor for a child in a remote village with no access to formal schooling. In healthcare, AI is already revolutionizing early diagnosis and treatment, bringing life-saving tools to underserved communities where doctors are few and far between—like AI-powered eye scans that detect, with remarkable accuracy, diabetic retinopathy in patrons of rural clinics.

Economically, AI is unlocking new pathways out of poverty: helping small farmers predict crop yields, enabling microloans for entrepreneurs without credit histories, and connecting local artisans to global markets through digital platforms. When inclusion is woven into the fabric of AI design, the technology does more than streamline systems—it uplifts individuals, unlocks human potential, and creates opportunities where none existed before. In this form, AI becomes not merely a driver of progress, but a powerful instrument of equity and justice.

Ethical AI Design and Governance

The foundation of a just AI future lies not only in the capabilities of the technology itself but in the values embedded in its design and governance. Ethical AI development begins with transparency—ensuring that algorithms are not black boxes, but systems whose logic and outcomes can be understood, questioned, and improved. This includes making models explainable to users and stakeholders, especially in high-stakes domains like healthcare, finance, and criminal justice.

Equally important is inclusivity: AI must be trained on diverse, representative datasets to avoid perpetuating historical injustices or marginalizing vulnerable communities. For instance, facial recognition systems that perform poorly on darker skin tones highlight the dangers of biased data.

To uphold public trust, accountability mechanisms must be built into every stage of the AI lifecycle. These include independent audits, algorithmic impact assessments, and enforceable regulatory frameworks that ensure AI systems are aligned with societal values and legal norms. But ethical governance is not a one-time checklist, it is a dynamic, ongoing process that must evolve alongside technological advances and shifting social expectations. This requires sustained collaboration between developers, policymakers, ethicists, and civil society organizations, creating a shared responsibility for shaping AI that is not only powerful, but principled. Only through this collective effort can we ensure that AI serves the public good and strengthens, rather than undermines, democratic values.

Human-AI Collaboration

Rather than replacing humans, artificial intelligence should be developed as a powerful ally to augment human capabilities, one that amplifies our abilities, sharpens our insights, and unlocks new realms of creativity and productivity.

Here are some well-known use cases of AI: In medicine, AI can act as a tireless assistant, rapidly analyzing complex medical data to help doctors diagnose conditions earlier and more accurately. For example, Google's DeepMind has developed an AI system that can detect over 50 eye diseases from retinal scans with accuracy comparable to world-leading experts, allowing ophthalmologists to focus more on patient care. In journalism, AI can comb through mountains of information—documents, social media, financial records—surfacing patterns and leads that fuel deeper, more impactful and accurate investigative reporting.

A notable case is The Washington Post's AI tool "Heliograf," which automatically generated thousands of articles on the 2016 US elections and the Rio Olympics, freeing up reporters to pursue in-depth stories. In the arts, AI is emerging as a creative collaborator, helping musicians compose, visual artists experiment, and writers explore new narrative forms. Microsoft's AI-powered Copilot in tools like Word and Excel helps users draft documents, analyze data, and automate repetitive tasks. This vision of human-AI collaboration honors human agency and ingenuity, blending the precision of machines with the empathy, intuition, and ethical judgment that only people can bring. It paints a future where AI doesn't compete but works alongside us—enhancing what we do best.

Global Cooperation and Regulation

The ethical development of AI cannot be confined within national borders. As AI systems increasingly influence global markets, politics, and cultures, international cooperation becomes essential. Frameworks like the OECD AI Principles and UNESCO's AI Ethics Recommendations provide a foundation for shared values and standards. These efforts promote responsible innovation while preventing a race to the bottom in terms of safety, privacy, and fairness. Global regulation also helps address cross-border challenges such as algorithmic bias, misinformation, exploitation of personal data by surveillance technologies, and autonomous weapons. By working together, nations can ensure that AI serves humanity as a whole, not just the interests of a few.

Dystopia or Design Flaw: Exploring the Unintended Consequences of Unchecked AI Development

Surveillance, Control, and Loss of Privacy

In a dystopian AI future, surveillance technologies become tools of control rather than protection. Authoritarian regimes can deploy facial recognition, predictive policing, and mass data collection to monitor citizens, suppress dissent, and manipulate behavior.

Even in democratic societies, the unchecked use of AI by corporations and governments can lead to the erosion of privacy and civil liberties. When individuals are constantly tracked, profiled, and targeted—often without their knowledge or consent—freedom of expression and autonomy are compromised. The ethical failure lies not in the technology itself, but in its deployment without transparency, accountability, or respect for human rights.

Bias, Discrimination, and Inequality

AI systems trained on biased data can perpetuate and even amplify existing social inequalities. From hiring algorithms that favor certain demographics to facial recognition systems that misidentify people of color, algorithmic injustice is a growing concern. These biases are often embedded in historical data and go unnoticed until they cause real harm—denying people jobs, loans, or justice. Worse, the opacity of many AI systems makes it difficult to detect or challenge these outcomes; hence we need more transparent AI systems. Without deliberate efforts to ensure fairness and inclusivity, AI risks becoming a digital mirror of systemic discrimination, reinforcing the very divides it could help bridge.

Autonomy and Job Displacement

As AI automates more tasks, from manufacturing to customer service to creative work, millions of jobs are at risk of being displaced. While some argue that new roles will emerge, the transition may not be equitable or timely, leaving many workers behind. Beyond economics, there is a deeper ethical concern: the erosion of human agency. When decisions about hiring, healthcare, or legal outcomes are made by cloudy algorithms, individuals lose control over their lives. The challenge is not just technological—it's moral. We must ask: Who benefits from automation, and who bears the cost?

Weaponization and Existential Risk

Perhaps the most alarming dystopian scenario involves the weaponization of AI. Autonomous drones, cyberweapons, and AI-driven misinformation campaigns pose serious threats to global security and democratic stability. Autonomous weapons can make life-and-death decisions without human intervention. This raises ethical concerns about accountability—who is responsible if the system makes a mistake? These technologies can be deployed at scale, with speed and precision that outpaces human oversight. Even more concerning is the possibility of creating AI systems whose goals diverge from human values, leading to unintended, potentially catastrophic consequences.

A strong source supporting the concerns about AI weaponization and existential risk comes from a 2024 statement signed by hundreds of AI researchers, which declared: "Mitigating the risk of extinction from AI should be a global priority alongside other societal-scale risks such as pandemics and nuclear war." This reflects a growing consensus among experts that the misuse of AI—through autonomous weapons, cyberwarfare, or systems misaligned with human values—poses real and urgent threats to global stability and human survival. There is currently no comprehensive global treaty governing the development and use of autonomous weapons, making it difficult to enforce ethical standards or prevent misuse.

The existential risk is not science fiction; it is a real and urgent concern among leading AI researchers. Without strict international safeguards, the misuse of AI could escalate conflicts, destabilize societies, and endanger humanity itself.

The Ethical Crossroads: What Determines the Path?

The Role of Policy and Regulation

At the heart of ethical AI development lies the framework of policy and regulation. Governments and international bodies play a critical role in setting the boundaries within which AI technologies can operate. These policies not only define what is legally permissible but also shape the incentives and constraints that guide innovation.

Effective regulation must strike a balance, encouraging technological advancement while safeguarding public interest. As AI systems become more autonomous and integrated into daily life, the urgency for adaptive, forward-thinking legislation grows. Regulatory frameworks must evolve in tandem with technological progress to ensure that ethical considerations remain central to AI deployment.

Laws, Standards, and Ethical Oversight

Beyond broad policy, the implementation of specific laws, technical standards, and ethical oversight mechanisms is essential. These tools provide the operational backbone for ensuring accountability and transparency in AI systems. Laws can enforce data protection, prevent algorithmic bias, and mandate explainability while standards help harmonize practices across industries and borders. Ethical oversight, through independent review boards, audits, and impact assessments, adds a layer of scrutiny that can catch unintended consequences before they cause harm. Together, these elements form a multi-layered defense against the misuse or unintended fallout of AI technologies.

Corporate Responsibility and Innovation Ethics

Tech companies are not just creators of AI, they are stewards of its societal impact, wielding immense power to shape the future. With their vast resources, global reach, and influence over public life, these corporations bear a profound responsibility to embed ethical considerations into every stage of AI development, from the earliest design decisions to real-world deployment. This means more than just compliance; it requires a proactive commitment to building systems that are fair, transparent, and accountable. It involves assembling diverse teams that can anticipate a wide range of social impacts, rigorously testing for bias and unintended consequences, and being open about how data is collected, used, and protected.

But ethical AI is not just about technical safeguards, it's about values. Innovation ethics demands that companies look beyond quarterly profits and consider the long-term consequences of their technologies: how they might reshape labor markets, influence democratic processes, or erode privacy norms. When AI systems are deployed without adequate oversight or reflection, the costs are borne not by the companies, but by the public, especially the most vulnerable. From algorithmic discrimination in decision-making such as hiring, to surveillance tools that threaten civil liberties, the risks are real and growing.

Ethical leadership in the private sector can set powerful precedents, influencing industry norms and public policy alike. Companies that lead with integrity can help build a future where AI enhances human well-being rather than undermines it. Because when innovation outpaces ethics, the question is no longer whether harm will occur, but who pays the price.

Public Awareness and Civic Engagement

An informed and engaged public is essential to shaping the ethical trajectory of AI. As these technologies increasingly affect everyday life, from making critical decisions to healthcare access, citizens must be equipped to understand and question how AI is used. Public education initiatives, accessible media coverage, and open forums for dialogue can empower individuals to participate in AI governance. Civic engagement also means holding institutions accountable and advocating for inclusive, equitable AI policies. When society at large is involved in the conversation, the path forward becomes more democratic, transparent, and aligned with shared human values.

Conclusion: Choosing the Future We Want

As we stand at the crossroads of technological evolution, the future of artificial intelligence remains unwritten. It holds the power to uplift humanity, curing diseases, democratizing education, and solving global challenges. Yet, it also carries the risk of deepening inequality, eroding privacy, and undermining human agency. Is a dark AI future foreseeable, or will all the decisions with regards to AI align towards utopian bliss and lead to a smarter, more integrated world of augmented humans with capabilities that are unmatched historically? This lies in the decisions we make with regards to how we handle AI.

This dual potential demands more than innovation—it demands ethical foresight. The choices we make today, from how we design algorithms to how we govern their use, will shape the world of tomorrow. It is not enough to ask what AI *can* do; we must ask what it *should* do. The future of AI depends on the conversations we have today.

The path to a utopian future is not automatic. It requires collective action, from policymakers, technologists, educators, and citizens alike. We must build systems that reflect our highest values: fairness, transparency, dignity, and justice. Human-AI collaboration represents a vision of the future where artificial intelligence is not a replacement for human intelligence, but a powerful partner that enhances our abilities. Rather than automating people out of the picture, AI should be designed to augment human strengths, supporting better decision-making, unlocking new levels of creativity, and boosting productivity across fields.

Ethics is not optional. It is the compass that must guide AI's journey. Only by embedding ethical principles at the heart of AI development can we ensure that this powerful technology serves the common good and leads us toward a future worth striving for. Governance and accountability can steer AI away from harm.

This collaborative model respects human agency and recognizes that the most impactful innovations come from the synergy between human intuition and machine precision. It also opens up new possibilities for accessibility: helping people with disabilities communicate, create, and work in ways that were previously out of reach.

Ultimately, a future built on human-AI collaboration is one where technology amplifies what makes us human: our empathy, imagination, and capacity for complex judgment. It's not about choosing between humans or machines; it's about designing systems where both thrive together. The power lies in our hands to build a world where AI helps us achieve a better future—by ensuring we develop safe, ethical systems that not only follow responsible policies but also evolve to serve humanity with fairness, transparency, and accountability.

About the Author

Sakina Syed is a dynamic senior data and AI consultant, AI engineer, and enthusiast with a stellar track record in the computer software industry, including notable tenures at tech giants like Microsoft. As a Microsoft AI engineer and Azure-certified professional, she excels in AI deployment, sales, management, teamwork, and leadership.

With a Bachelor of Science in Neuroscience and Mental Health Studies from the University of Toronto, Sakina has also enriched her expertise with courses in business management and IT. Her passion for writing and traveling adds a unique dimension to her professional persona.

Sakina's enthusiasm for business and the application of business psychology to understand consumer needs is matched by her extensive seven-year experience in customer service. She is adept at interpreting consumer feedback and statistical data, enabling her to identify market requirements with precision. Her strong passion for team management and leadership drives her dedication to delivering exceptional customer experiences and fostering business success.

Follow her here!

LinkedIn: https://www.linkedin.com/in/sakina-syed/

Website: www.youraiconsultant.ca

Email: info@youraiconsultant.ca

CHAPTER 33

AI SECURITY AND SAFETY: NAVIGATING THE FUTURE WITH CURIOSITY AND CAUTION

By Ersin Uzun, PhD
Professor, Innovator, Entrepreneur
Pittsford, New York

Technology is neither good nor bad; nor is it neutral.
—Melvin Kranzberg, Historian of Technology

Artificial intelligence (AI) isn't a futuristic concept waiting to change the world—it's already here. It shapes the music you listen to, powers your email's spam filter, and gives you recommendations on what to watch on Netflix or read on Google News; it even helps doctors interpret X-ray images. Large language models (LLMs) such as ChatGPT, Llama, or Gemini can prepare slides or reports, teach you a new subject, or write software. If you used a smartphone, relied on navigation apps, searched something on Google, or had a conversation with ChatGPT, you've already interacted with AI more than once today.

These technologies offer convenience and increase productivity, but they might come with hidden costs and risks. The same algorithms that curate your social media feed or recommend news articles can prevent you from getting diverse views, create echo chambers, promote deception or even radicalization. AI-powered enterprise software can speed up sorting through applicant resumes or make loan decisions faster but may also introduce or increase bias. Deepfake videos and AI-generated voice clones can be entertaining but are also used to mislead, defraud, or manipulate people. LLMs can be a great tutor to learn a new subject, but step-by-step instructions and malware produced by dark LLMs (malicious versions LLMs found in dark web or hacker forums) can turn anyone into a capable cyber-criminal within a few hours. AI impacts your personal and professional life, whether you're aware of it or not. And as its influence deepens, understanding its promises and pitfalls become a matter of personal empowerment and civic responsibility.

We Are on the Brink of the AI Revolution

At its core, artificial intelligence is about machines performing tasks that typically require human intelligence—such as recognizing speech, interpreting images, making decisions, or learning from data. The reality of AI today is more expansive than many realize.

Corporations are rapidly adopting AI for everything from fraud detection and logistics planning to making decisions related to human resources or marketing. AI chatbots field customer service queries, predictive algorithms steer inventory management, and automated systems make hiring decisions.

For individuals, AI applications have also exploded. Tools like ChatGPT can assist with drafting emails, summarizing complex information, creating presentations, generating code, writing articles, and even planning travel itineraries. AI art generators produce stunning images and videos. Autonomous vehicles can drive people around without a driver. Most apps on your phone or devices in your house continuously learn your habits and preferences to elevate your experience and comfort. These advances offer significant benefits, making life easier with levels of automation and productivity gains that were not possible before.

On the other hand, the dark side of AI is real, and it is important to be aware of it. Unchecked and unregulated AI development and deployment can introduce biases leading to unfair or unethical processes, create unprecedented levels of cybersecurity risks, and lead to overreliance on a technology that could make decisions without any common sense, values, or ethics. As we delegate more decisions to AI algorithms, we risk losing transparency, accountability, and human empathy.

Understanding the strengths and weaknesses of AI helps you harness its power while guarding against its flaws; you don't need to be a computer scientist but still need to be informed and thoughtful. Like electricity, the internet, or mobile phones, AI will be an indispensable part of modern life, but it will likely change our lives more profoundly and rapidly than any of its predecessors did.

Why Cybersecurity Is a Growing Concern with AI Adoption?

As we automate more systems—from autonomous vehicles to critical infrastructure—the potential points of failure and consequences of such failures expand dramatically. By now, you might be used to getting notices about a company losing your personal information to hackers, intermittent service disruptions of online services due to cyberattacks, and even dealing with fraudulent charges or identity theft attempts by cyber criminals. These incidents are frustrating, inconvenient, and might occasionally cause financial losses. With the rapid adoption of AI and automation at every pillar of modern life, the future cybersecurity failures, however, are undoubtedly scarier.

Consider autonomous vehicles: Driverless vehicles are already offering ride-hauling services in a handful of major cities like San

Francisco, Phoenix, and Los Angeles. It is only a matter of time before we see autonomous cars, trucks, buses, and construction/agricultural machinery everywhere. What if they are hacked and start to drive into crowds, schools, or playgrounds? Financial systems increasingly use AI for high-frequency trading. What happens when a cyberattack sends markets crashing? Smart cities rely on AI for traffic flow, mass transportation, surveillance, and utilities. A single compromise could trigger cascading failures, leaving millions stranded, without power or in grave danger.

AI democratizes the power to attack in cyberspace. Novice hackers can now deploy sophisticated phishing campaigns or generate malware with AI. With the introduction of intelligent ransomware-as-a-service platforms, malicious chatbots, or deepfake tools that can imitate anyone on a video or phone call, the barrier to commit cybercrime is reduced to a computer with internet access and lack of a sound moral compass.

Misinformation is yet another real threat that is frequently downplayed by people that might see it as a tool to influence public opinion or technology companies that want to avoid accountability and the cost of developing safeguards against it. AI-generated content can flood social media with false narratives at scale, which in return can erode trust in institutions, influence elections, and radicalize or polarize people to incite violence.

These aren't distant hypotheticals; many of these risks have already surfaced and unfortunately are continuously growing. AI doesn't just raise the stakes like faster internet or more powerful computers did; it changes the game entirely. Our digital ecosystem is getting interconnected to every facet of modern life, and we are more vulnerable than ever before. The consequences of security and safety failures in an AI-driven future will not be just inconveniences or financial loss, but have the real potential to cost lives, cripple whole communities, and erode the civil norms of society.

AI Is Neither Trustworthy nor Safe, Not Yet

Today's AI technology is mostly a black box. Even the researchers that have created the leading AI models with billions of parameters today cannot truly understand or explain how and why it works in certain cases, nor can they predict when it would or could make a mistake.

Consider large language models (LLMs), like ChatGPT. They are nothing short of impressive in answering questions on any topic or handling tasks that used to require experts with years of education and training. However, it is crucial to understand these models are far from being an oracle of truth. In the simplest of terms, LLMs are just algorithms that probabilistically construct sentences word by word and do so by predicting the next word based on the words that came before.

That prediction is based on the patterns observed in huge amounts of text (books, websites, articles, etc.) used in their training. Patterns observed in historic data enables LLMs to create new plausible patterns, but doing so with no intrinsic ability to differentiate good from bad or true from false. That is why these LLMs sometimes make up references to non-existing sources, might give examples or information that is factually wrong or fictional, fail at simple logic problems, or simply learn the wrong thing from the patterns that exist in their training data (e.g., an LLM might learn the historical pattern of people of certain gender or race being hired less in the past and reproduce the same pattern in its decisions, effectively basing the hiring decisions on gender and race rather than the qualifications of applicants). In other words, what we call AI today doesn't have the common sense, sound logical reasoning or ethical values we would expect from our human peers; it just learns the patterns in its training data and tries to replicate them without understanding or questioning what might have caused those patterns or whether we want to really replicate them. A recent AI model passing a multistate bar exam should not give anyone confidence that it understands law, logic, or is capable of always giving correct information.

You might think that when a chatbox like ChatGPT makes mistakes, it is not a big deal, but this belief is incomplete, deceptive, and dangerous. Corporations and even governments that are eager to reduce costs, speed up their processes, or be the first to introduce new products or services are adopting these same AI models at a surprising pace. Your next application for a job, permit, loan, or visa might be evaluated by one of these AI models in the backend without you even realizing. The article, blog post, or novel you read; the screenplay of a recent movie or TV show you watch might have already been largely generated by LLMs. Even the diagnoses report on your next X-ray or ultrasound might be written by AI rather than a doctor.

Today's AI is a class of algorithms that can learn patterns in the data they were given and be either used to generate similar patterns or decide whether a pattern is similar to one learned before. This makes AI itself vulnerable to a class of attacks or malicious manipulations that we call data or model poisoning. Let's consider an oversimplified example with an AI model trained to make loan decisions again. If the AI model's training data shows all previous loans were approved for people that were born on Feb 29th, then the AI will likely learn that pattern and might approve a future loan if the applicant's birthday is Feb 29th without taking other factors into account. That pattern might be there by coincidence or due to malicious manipulation of its training data.

The reality of AI models being enormous black boxes with billions of parameters makes it extremely hard to detect such backdoors (i.e., pre-programmed malicious behaviors on certain inputs) or biases it might have. We currently don't have the technology to be able to validate any complex AI model to be free of such manipulations or biases. If you extrapolate this simple example to autonomous cars, planes, factories, drones, security systems, and pretty much everything else you can imagine, it is not hard to envision the potential danger posed.

If you are thinking that corporations or governments would behave responsibly and would never adopt something that is potentially unsafe or secure, I suggest you take a closer look at the history of thousands of lawsuits, settlements, recalls, insurance claims, and decades of research that surfaced many unsafe products and unfair practices by corporations or government agencies. Like everything else, you are ultimately responsible for the well-being of yourself and your business, so be curious and explore how you can harness the power of AI but also make sure to be informed about the pitfalls that come with it and how you can safeguard against them.

Principles for Responsible AI Innovation and Adoption

With power comes responsibility. As individuals, businesses, and governments rush to adopt AI, it's essential to do so thoughtfully. I believe there are five foundational principles that we should keep ourselves, businesses, and elected officials accountable for:

1. *Transparency*: AI systems should be understandable and explainable. Anything less should not be given autonomous decision-making power. If an algorithm denies your loan, you deserve to know why. Black-box models—where even developers can't explain decisions—undermine trust and accountability.
2. *Safety and security*: AI and AI-driven systems must be resilient against misuse and attacks. That means designing systems that can withstand cyberattacks, manipulation, or unintended behaviors. Safety testing and monitoring should be rigorous, continuous, and public. There should be clear liability and accountability to creators and operators of AI systems in case of security and safety failures.
3. *Fairness and equity*: AI must not replicate or deepen social inequalities. Training data and outcomes should be audited for bias. Fairness isn't a technical add-on—it's a moral imperative.
4. *Privacy*: AI systems often rely on large amounts of personal data. That data must be collected and used responsibly, with clear consent, robust protection, and respect to intellectual property. Privacy or respecting intellectual property isn't just about compliance—it's about dignity.

5. *Human oversight*: Critical decisions, from healthcare to criminal justice, should never be left entirely to AI, especially to the current generation of the technology. AI can assist, but humans must remain in control. Oversight ensures that empathy, context, and ethical judgment stay in the loop.

These principles aren't abstract ideals, they're the foundation of a safe, inclusive, and trustworthy AI future.

Be Curious. Be Cautious. Be a Force for Good

It's natural to feel overwhelmed by AI's complexity or worried about how profoundly it might change our lives. But the solution isn't panic or passivity—it's proactive curiosity and informed critical thinking. Make sure to continuously experiment with it to increase your AI literacy and understand its power, but do it safely, knowing its pitfalls and weaknesses. Read articles. Ask questions. Follow experts who challenge your views. If you're adopting AI in your workplace, demand transparency and accountability from vendors. If you're using AI tools at home, understand what data they collect and think about what might go wrong.

Be cautious but not paralyzed. AI is a tool, and like all tools, its impact depends on how we use it. Advocate for better policies and legislation. Support ethical innovation, criticize irresponsible adoptions. And most importantly, model responsible, curious, and informed behavior for others.

We are still in the early chapters of the AI era. The choices we make now—about ethics, safety, and security—will shape what comes next. AI will help us build a better world, but only if we guide it with wisdom, courage, and care.

About the Author

Ersin Uzun, PhD, is a computer scientist, entrepreneur, and an innovation consultant. He is currently the Katherine Johnson Endowed Executive Director of the Global Cybersecurity Institute and a professor of cybersecurity at Rochester Institute of Technology (RIT).

Before joining RIT, Ersin was the vice president of R&D and the head of new ventures at Palo Alto Research Center (PARC), and worked at Hewlett Packard Labs, Nokia Research Center, and the French National Institute for Research in Digital Science and Technology prior to that.

Ersin has over 150 global patents and 45 scientific publications. He is an inductee of Hall of Fame at the University of California, Irvine where he received his MS and PhD in Computer Science, and has served in advisory and review boards of US National Science Foundation, Finnish Funding Agency for Technology and Innovation, Research Council of Canada, and Scientific and Technological Research Council of Turkey, as well as on the boards and advisory boards of startups and academic institutions. He is an experienced speaker, frequently giving talks on the topics of cybersecurity, AI, technology development, and innovation management.

Email: ersin@ersinuzun.com

Website: www.ersinuzun.com

DID YOU ENJOY THIS BOOK?

If you enjoyed reading this book, you can help by suggesting it to someone else you think might like it, and **please leave a positive review** wherever you purchased it. This does a lot in helping others find the book. We thank you in advance for taking a few moments to do this.

THANK YOU

You might also like other Thin Leaf Press titles:

The AI Revolution: Thriving Within Civilization's Next Big Disruption

The AI Mindset: Thriving Within Civilization's Next Big Disruption

AI: Work Smarter and Live Better Within Civilization's Next Big Disruption

Peak Performance: Mindset Tools for Managers

Peak Performance: Mindset Tools for Sales

Peak Performance: Mindset Tools for Leaders

Peak Performance: Mindset Tools for Business

Peak Performance: Mindset Tools for Entrepreneurs

Peak Performance: Mindset Tools for Athletes

The Successful Mind: Tools to Living a Purposeful, Productive, and Happy Life

The Successful Body: Using Fitness, Nutrition, and Mindset to Live Better

The Successful Spirit: Top Performers Share Secrets to a Winning Mindset

Winning Mindset: Elite Strategies for Peak Performance

Winner's Mindset: Peak Performance Strategies for Success

The Life Coach's Tool Kit, Vol. 1

The Life Coach's Tool Kit, Vol. 2

The Life Coach's Tool Kit, Vol. 3

Ordinary to Extraordinary

The Magical Lightness of Being

Explore.

www.ingramcontent.com/pod-product-compliance
Lightning Source LLC
Chambersburg PA
CBHW071317210326
41597CB00015B/1256

9781968318086